More praise for

"A worthy addition to the Fe
follow-up to the standard-bearer, James Gleick's *Genius*."
—*Kirkus Reviews*

"Enlightening." —George Johnson, *New York Times*

"Entertaining and masterly. A great read."
—Brian Greene, author of *The Elegant Universe*

"Such a charismatic figure deserves a charismatic, knowledgeable, and literate physicist as his warts-and-all biographer. Lawrence Krauss fits the bill admirably and rises to the challenge with style, panache, and deep understanding."
—Richard Dawkins, author of *The God Delusion*

"Krauss's wonderful biography puts Feynman's remarkable contributions to science front and center, accessibly, in the context of his life and times. Feynman would approve."
—Frank Wilczek, MIT, Nobel Laureate in Physics

"Highly recommended for readers who want to get to know one of the preeminent scientists of the 20th century."
—*Publishers Weekly*

"A rich and entertaining biography." —Dan Falk, *New Scientist*

"If your interest is in Feynman the physicist, [*Quantum Man*] is an excellent place to start." —Jon Turney, *Times Higher Education*

"An enlightening addition to the field." —George Johnson, *The Scotsman*

PUBLISHED TITLES IN THE GREAT DISCOVERIES SERIES

Sherwin B. Nuland
The Doctors' Plague: Germs, Childbed Fever,
and the Strange Story of Ignác Semmelweis

Michio Kaku
Einstein's Cosmos: How Albert Einstein's Vision Transformed
Our Understanding of Space and Time

Barbara Goldsmith
Obsessive Genius: The Inner World of Marie Curie

Rebecca Goldstein
Incompleteness: The Proof and Paradox of Kurt Gödel

Madison Smartt Bell
Lavoisier in Year One: The Birth of a New Science
in an Age of Revolution

George Johnson
Miss Leavitt's Stars: The Untold Story of the Woman
Who Discovered How to Measure the Universe

David Leavitt
The Man Who Knew Too Much: Alan Turing
and the Invention of the Computer

William T. Vollmann
Uncentering the Earth: Copernicus
and The Revolutions of the Heavenly Spheres

David Quammen
The Reluctant Mr. Darwin: An Intimate Portrait of Charles Darwin
and the Making of His Theory of Evolution

Richard Reeves
A Force of Nature: The Frontier Genius of Ernest Rutherford

Michael Lemonick
The Georgian Star: How William and Caroline Herschel Revolutionized Our Understanding of the Cosmos

Dan Hofstadter
The Earth Moves: Galileo and the Roman Inquisition

ALSO BY LAWRENCE M. KRAUSS

Hiding in the Mirror

Atom

Quintessence

Beyond Star Trek

The Physics of Star Trek

Fear of Physics

The Fifth Essence

LAWRENCE M. KRAUSS

Quantum Man

Richard Feynman's Life in Science

with corrections by Cormac McCarthy

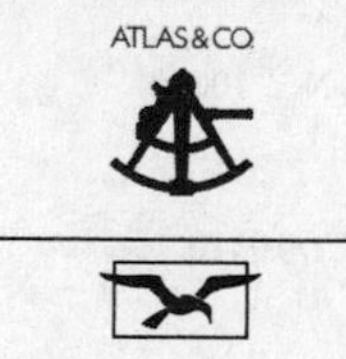

W. W. NORTON & COMPANY
NEW YORK · LONDON

Printed in the United States of America
First published as a Norton paperback 2012

For information about permission to reproduce selections from this book, write to Permissions, W. W. Norton & Company, Inc., 500 Fifth Avenue, New York, NY 10110

For information about special discounts for bulk purchases, please contact W. W. Norton Special Sales at specialsales@wwnorton.com or 800-233-4830

Manufacturing by Courier Westford
Production manager: Anna Oler

Library of Congress has catalogued the hardcover edition as follows:
Krauss, Lawrence Maxwell.
Quantum man : Richard Feynman's life in science /
Lawrence M. Krauss. — 1st ed.
p. cm. — (Great discoveries)
Includes bibliographical references and index.
ISBN 978-0-393-06471-1 (hardcover)
1. Feynman, Richard P. (Richard Phillips), 1918–1988.
2. Physicists—United States—Biography. I. Title. II. Series.
QC16.F49K73 2011
530.092—dc22
[B]

2010045512

ISBN 978-0-393-34065-5 pbk.

Atlas & Co.
15 West 26th Street, New York, N.Y. 10010

W. W. Norton & Company, Inc.
500 Fifth Avenue, New York, N.Y. 10110
www.wwnorton.com

W. W. Norton & Company Ltd.
Castle House, 75/76 Wells Street, London W1T 3QT

2 3 4 5 6 7 8 9 0

Reality must take precedence over public relations, for nature cannot be fooled.

—RICHARD P. FEYNMAN, 1918–1988

Contents

Introduction

> I find physics is a wonderful subject. We know so very much and then subsume it into so very few equations that we can say we know very little.
>
> —Richard Feynman, 1947

It is often hard to disentangle reality from imagination when it comes to childhood memories, but I have a distinct recollection of the first time I thought that being a physicist might actually be exciting. As a child I had been fascinated with science, but the science I had studied was always removed from me by at least a half century, and thus it hovered very close to history. The fact that not all of nature's mysteries had been solved was not yet firmly planted in my mind.

The epiphany occurred while I was attending a high school summer program on science. I don't know if I appeared bored or not, but my teacher, following our regularly scheduled lesson, gave me a book titled *The Character of Physical Law* by Richard Feynman and told me to read the chapter on the distinction between past and future. It was my first contact with the notion of entropy and disorder, and like many people before me, including the great physicists Ludwig Boltzmann and Paul Ehrenfest, who killed themselves after devoting much of their careers to developing this subject, it left me befuddled and frustrated. How the world changes as one goes from considering simple problems involving two objects, like the earth and the moon, to a system involving many particles, like the gas molecules in the room in which I am typing this, is both subtle and profound—no doubt too subtle and profound for me to appreciate at the time.

But then, the next day, my teacher asked me if I had ever heard of antimatter, and he proceeded to tell me that this same guy Feynman had recently won the Nobel Prize because he explained how an antiparticle could be thought of as a particle going backward in time. Now that really fascinated me, although I didn't understand any of the details (and in retrospect I realize my teacher didn't either). But the notion that these kinds of discoveries were happening during my lifetime inspired me to think that there was a lot left to explore. (Actually while my conclusion was true, the information that led to it wasn't. Feynman had published his Nobel Prize–winning work on quantum electrodynamics almost a decade before I was born, and the ancillary idea that antiparticles could be thought of as particles going backward in time wasn't even his. Alas, by the time ideas filter down to high school teachers and texts, the physics is usually twenty-five to thirty years old, and sometimes not quite right.)

As I went on to study physics, Feynman became for me, as he did for an entire generation, a hero and a legend. I bought his *Feynman Lectures on Physics* when I entered college, as did most other aspiring young physicists, even though I never actually took a course in which these books were used. But also like most of my peers, I continued to turn to them long after I had moved on from the so-called introductory course in physics on which his books were based. It was while reading these books that I discovered how my summer experience was oddly reminiscent of a similar singular experience that Feynman had had in high school. More about that later. For now I will just say that I only wish the results in my case had been as significant.

It was probably not until graduate school that I fully began to understand the ramifications of what that science teacher had been trying to relate to me, but my fascination with the world of fundamental particles, and the world of this interesting guy Feynman, who wrote about it, began that summer morning in high school and in large part has never stopped. I just remembered, as I was writing this, that I chose to write my senior thesis on path integrals, the subject Feynman pioneered.

Through a simple twist of fate, I was fortunate enough to meet and spend time with Richard Feynman while I was still an undergraduate. At the time I was involved with an organization called the Canadian Undergraduate Physics Association, whose sole purpose was to organize a nationwide conference during which distinguished physicists gave lectures and undergraduates presented results from their summer research projects. It was in 1974, I think, that Feynman had been induced (or seduced, I don't know and shouldn't presume) by the very attractive president of the organization to be the keynote speaker at that year's conference in Vancouver. At the meeting I had the temerity to ask him a question after his lecture, and a photographer from a national magazine took a picture of the moment and used the photo, but more important, I had brought my girlfriend along with me, and one thing led to another and Feynman spent much of the weekend hanging out with the two of us in some local bars.

Later, while I was at graduate school at MIT, I heard Feynman lecture several times. Years later still, after I had received my PhD and moved to Harvard, I presented a colloquium at Caltech, and Feynman was in the audience, which was slightly unnerving. He politely asked a question or two and then came up afterward to continue the discussion. I expect he had no memory of our meeting in Vancouver, and I am forever regretful of the fact that I never found out, because while he waited patiently to talk to me, a persistent and rather annoying young assistant professor monopolized the discussion until Feynman finally walked off. I never saw him again, as he died a few years later.

RICHARD FEYNMAN WAS a legend for a whole generation of physicists long before anyone in the public knew who he was. Getting a Nobel Prize may have put him on the front page of newspapers around the world, but the next day there are new headlines, and any popular name recognition usually lasts about as long as the newspaper itself. Feynman's popular fame thus did not arise from his scientific discoveries, but began through a series of books recounting his personal reminiscences. Feynman the raconteur was every bit as creative and fascinating as Feynman the physicist. Anyone who came into personal contact with him had to be struck immediately by his wealth of charisma. His piercing eyes, impish smile, and New York accent combined to produce the very antithesis of a stereotypical scientist, and his personal fascination with such things as bongo drums and strip bars only added to his mystique.

As often happens however, the real catalyst that made Feynman a public figure arose by accident, in this case a tragic accident: the explosion shortly after liftoff of the *Challenger* space shuttle, which was carrying the first "civilian," a public school teacher who was scheduled to teach some classes from space. During the investigation that ensued, Feynman was asked to join the NASA investigatory panel, and in an uncharacteristic moment (he studiously avoided committees and anything else that kept him away from his work), he agreed.

Feynman pursued the task in his own, equally uncharacteristic way. Rather than study reports and focus on bureaucratic proposals for the future, Feynman talked directly to the engineers and scientists at NASA, and in a famous moment during the televised hearings, he performed an experiment, putting a small rubber O-ring in a glass of ice water and thus demonstrating that the O-rings used to seal the rocket could fail under temperatures as cold as those on the day of the ill-fated launch.

Since that day, books chronicling his reminiscences, compilations of his letters, audiotapes of "lost lectures," and so on, have appeared, and following his death, his legend has continued to grow. Popular Feynman biographies have also been published, with the most notable being James Gleick's masterful *Genius.*

Feynman the human being will always remain fascinating, but when I was approached about producing a short and accessible volume that might reflect Feynman the man as seen through his scientific contributions, I couldn't resist. The exercise motivated me because I would be reviewing all of his original papers. (Most people may not realize that it is rare for scientists to go back to the original literature in their field, especially if the work is more than a generation old. Scientific ideas get distilled and refined, and most modern presentations of the same physics often bear very little resemblance to the initial formulations.) But more important, I realized that Feynman's physics provides, in microcosm, a perspective on the key developments in physics over the second half of the twentieth century, and many of the puzzles he left unresolved remain with us today.

In what follows I have tried to do justice to both the letter and the spirit of Feynman's work in a way in which he might have approved. Perhaps for this reason this book is first and foremost about Feynman's impact on our current understanding of nature, as reflected within the context of a personal scientific biography. I will devote little space to the many arcane blind alleys and red herrings that lure even the most successful scientists—and Feynman was no exception—as they claw their way to scientific understanding. It is hard enough, without having to sort through these false starts, for nonexperts to gain a proper perspective of what physicists have learned about the natural world. No matter how elegant or brilliant some of the false starts may be, ultimately what matters are the ideas that have survived by satisfying the test of experiment.

My modest goal therefore is to focus on Feynman's scientific legacy as it has affected the revolutionary discoveries of twentieth-century physics, and as it may impact any unraveling of the mysteries of the twenty-first century. The insight I really want to reveal to nonphysicists, if I can, is why Feynman has reached the status of a mythic hero to most physicists now alive on the planet. If I can capture that, I will have helped readers understand something central about modern physics and Feynman's role in changing our picture of the world. That, to me, is the best testimony I can give to the genius that was Richard Feynman.

PART I

The Paths to Greatness

Science is a way to teach how something gets to be known, what is not known, to what extent things are known (for nothing is known absolutely), how to handle doubt and uncertainty, what the rules of evidence are, how to think about things so that judgments can be made, how to distinguish truth from fraud, and from show.

—Richard Feynman

CHAPTER 1

Lights, Camera, Action

> Perhaps a thing is simple if you can describe it fully in several different ways without immediately knowing that you are describing the same thing.
>
> —RICHARD FEYNMAN

Could one have guessed while he was still a child that Richard Feynman would become perhaps the greatest, and probably the most beloved, physicist of the last half of the twentieth century? It is not so clear, even if many of the incipient signs were there. He was undeniably smart. He had a nurturing father who entertained him with puzzles and instilled a love of learning, encouraging his innate curiosity and feeding his mind whenever possible. And he had a chemistry set and displayed a fascination with radios.

But these things were not that uncommon for bright youngsters at the time. In most fundamental respects Richard Feynman appeared to be a typical smart Jewish kid from Queens growing up after the First World War, and it is perhaps that simple fact as much as anything else that colored his future place in history. His mind was extraordinary, yes, but he remained firmly grounded in reality, even as he was driven to explore the most esoteric realms of our existence. His disrespect for pomposity came from an early life in which he was not exposed to it, and his disrespect for authority came not only from a father who nurtured this independence but also from an early life in which he was remarkably free to be a child, to follow his own passions, and to make his own mistakes.

Perhaps the first signal of what was to come was Feynman's literally indefatigable ability to concentrate on a problem for hours at a time, so much so that his parents began to worry. As a teenager, Feynman made practical use of his fascination with radios: he opened a small business fixing them. But unlike conventional repairmen, Feynman would delight in solving radio problems not merely by tinkering, but by thinking.

And he would combine this remarkable ability to focus all of his energy on a problem with an innate talent as a showman. His most famous radio repair, for example, involved an episode where he paced back and forth thinking while the broken radio shrieked in front of its owner whenever it was first turned on. Finally young Feynman pulled out two tubes and exchanged them, solving the problem. My suspicion is that Feynman let the whole thing last longer than it needed to, just for effect.

In later life almost exactly the same story would be told again. But this one originated when a skeptical Feynman was asked to examine a puzzling photograph from a bubble chamber—a device where elementary particles would leave visible tracks. After thinking for a while, he placed his pencil down on a precise spot in the picture and claimed that there must be a bolt located right there, where a particle had had an unanticipated collision, producing results that otherwise had been misinterpreted. Needless to say, when the experimenters involved in the claimed discovery went back to their device and looked at it, there was the bolt.

The showmanship, while contributing to the Feynman lore, was not important to his work however. Neither was his fascination with women, which also emerged later. The ability to concentrate, combined with an almost superhuman energy that he could apply to a problem, was. But the final essential icing on the cake, when combined with the former two characteristics, ultimately ensured his greatness. It involved simply an almost unparalleled talent for mathematics.

Feynman's mathematical genius began to manifest itself by the time he was in high school. While a sophomore he taught himself trigonometry, advanced algebra, infinite series, analytical geometry, and differential and integral calculus. And in his self-learning, the other aspect of what made Feynman so unique began to materialize: he would recast all knowledge in his own way, often inventing a new language or new formalism to reflect his own understanding. In certain cases necessity was the mother of invention. When typing out a manual on complex mathematics, in 1933, at the age of fifteen, he devised "typewriter symbols" to reflect the appropriate mathematical operations, since his typewriter did not have keys to represent them, and created a new notation for a table of integrals that he had developed.

Feynman entered MIT with the intent to study mathematics, but it was a misplaced notion. Even though he loved mathematics, he forever wanted to know what he could "do" with it. He asked the chairman of the mathematics department this question and got two different answers: "Insurance estimates," and "If you have to ask that, then you don't belong in mathematics." Neither resonated with Feynman, who decided mathematics wasn't for him, so he switched to electrical engineering. Interestingly, this switch seemed too extreme. If mathematics was without purpose, engineering was too practical. Like the soup in the Goldilocks tale, however, physics was "just right," and by the end of his freshman year Feynman had become a physics major.

The choice of course was an inspired one. Feynman's innate talents allowed him to excel in physics. But he had another talent that mattered even more perhaps, and I don't know if it was innate or not. This was intuition.

Physical intuition is a fascinating, ephemeral kind of skill. How does one know which avenue of approach will be most fruitful to solve a physics problem? No doubt some aspects of intuition are acquired. This is why physics majors are required to solve so many problems. In this way, they begin to learn which approaches work and which don't, and increase their toolkit of techniques along the way. But surely some aspect of physical intuition cannot be taught, one that resonates at a certain place and time. Einstein had such intuition, and it served him well for over twenty years, from his epochal work on special relativity to his crowning achievement, general relativity. But his intuition began to fail him as he slowly drifted away from the mainstream of interest in quantum mechanics in the twentieth century.

Feynman's intuition was unique in a different way. Whereas Einstein developed completely new theories about nature, Feynman explored existing ideas from a completely new and usually more fruitful perspective. The only way he could really understand physical ideas was to derive them using his own language. But because his language was usually also self-taught, the end results sometimes differed radically from what "conventional" wisdom produced. As we shall see, Feynman created his own wisdom.

But Feynman's intuition was also earned the hard way, based on relentless labor. His systematic approach and the thoroughness with which he examined problems were already evident in high school. He recorded his progress in notebooks, with tables of sines and cosines he had calculated himself, and later on in his comprehensive calculus notebook, titled "The Calculus for the Practical Man," with extensive tables of integrals, which again he had worked out himself. In later life he would amaze people by proposing a new way to solve a problem, or by grasping immediately the heart of a complex issue. More often than not this was because at some time, in the thousands of pages of notes he kept as he worked to understand nature, Feynman had thought about that very problem and explored not just one, but a host of different ways of solving it. It was this willingness to investigate a problem from every vantage point, and to carefully organize his thinking until he had exhausted all possibilities—a product of his deep intellect and his indefatigable ability to concentrate—that set him apart.

Perhaps *willingness* is the wrong word here. *Necessity* would be a better choice. Feynman needed to fully understand every problem he encountered by starting from scratch, solving it in his own way and often in several different ways. Later on, he would try to imbue this same ethic to his students, one of whom later said, "Feynman stressed creativity—which to him meant working things out from the beginning. He urged each of us to create his or her own universe of ideas, so that our products, even if only answers to assigned classwork problems, would have their own original character—just as his own work carried the unique stamp of his personality."

Not only was Feynman's ability to concentrate for long periods evident when he was young, but so was his ability to control and organize his thoughts. I remember having a chemistry set when I was a kid and I also remember often randomly throwing things together to see what would happen. But Feynman, as he later emphasized, "never played chaotically with scientific things." Rather he always carried out his scientific "play" in a controlled manner, always attentive to what was going on. Again, much later, after his death, it became clear from the extensive notes he took that he carefully recorded each of his explorations. He even considered at one point organizing his domestic life with his future wife along scientific lines, before a friend convinced him that he was being hopelessly unrealistic. Ultimately, his naivete in this regard disappeared, and much later he advised a student, "You cannot develop a personality with physics alone. The rest of life must be worked in." In any case, Feynman loved to play and joke, but when it came to science, starting early on and continuing for the rest of his life, Feynman could be deadly serious.

He may have waited until the end of his first year of university to declare himself a physics major, but the stars aligned when he was still in high school. In retrospect, what might have been the defining moment occurred when his high school teacher, Mr. Bader, introduced him to one of the most subtle and wonderful hidden mysteries of the observable world, a fact that had built on a discovery made three hundred years before he was born by a brilliant and reclusive lawyer-turned-mathematician, Pierre de Fermat.

Like Feynman, Fermat would achieve public recognition late in life for something that was unrelated to his most substantial accomplishments. In 1637, Fermat scrawled a brief note in the margin of his copy of *Arithmetica*, the masterpiece by the famous Greek mathematician Diophantus, indicating that he had discovered a simple proof of a remarkable fact. The equation $x^n + y^n = z^n$ has no integer solutions if $n>2$ (for $n=2$, this is familiar as the Pythagorean theorem relating lengths of the sides of a right triangle). It is doubtful that Fermat really possessed such a proof, which 350 years later required almost all of the developments of twentieth-century mathematics and several hundred pages to complete. Nevertheless, if Fermat is remembered at all today among the general public, it is not for his many key contributions to geometry, calculus, and number theory, but rather for this speculation in the margin that will forever be known as *Fermat's last theorem.*

Twenty-five years after making this dubious claim, Fermat did present a complete proof of something else, however: a remarkable and almost otherworldy principle that established an approach to physical phenomena that Feynman would use later to change the way we think about physics in the modern world. The issue to which Fermat turned his attention in 1662 involved a phenomenon the Dutch scientist Willebrord Snell had described forty years earlier. Snell discovered a mathematical regularity in the way light is bent, or refracted, when it crosses between two different media, such as air and water. Today we call this Snell's law, and it is often presented in high school physics classes as yet one additional tedious fact to be memorized, even though it played a profoundly important role in the history of science.

Snell's law pertains to the angles that a light ray makes when transmitted across the surface between two media. The exact form of the law is unimportant here; what is important is both its general character and its physical origin. In simple terms, the law states that when light goes from a less dense to a more dense medium, the trajectory of the light ray is bent closer to the perpendicular to the surface between the media (see figure).

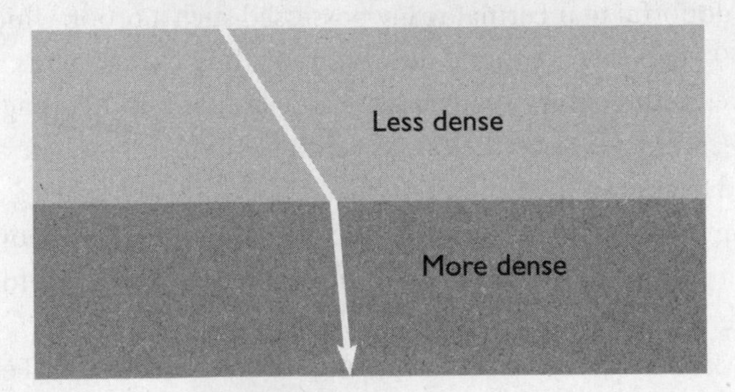

Snell's law

Now, why does the light bend? Well, if light were made up of a stream of particles, as Newton and others thought, one could understand this relationship if the particles speed up as they move from one medium to the other. They would literally be dragged forward, moving more effectively in a direction perpendicular to the surface they had just crossed. However, this explanation seemed fishy even at the time. After all, in a more dense medium any such particles would presumably encounter a greater resistance to their motion, just as cars on a road end up moving more slowly in heavy traffic.

There was another possibility, however, as the Dutch scientist Christiaan Huygens demonstrated in 1690. If light were a wave and not made of particles, then just as a sound wave bends inward when it slows down, the same would occur for light if it too slowed down in the denser medium. As anyone familiar with the history of physics knows, light does indeed slow down in denser media, so that Snell's law provides important evidence that light behaves, in this instance, like a wave.

Amost thirty years before Huygens's work, Fermat too reasoned that light should travel more slowly in dense media than in less dense media. Instead of thinking in terms of whether light was a wave or particle, however, Fermat the mathematician showed that in this case one could explain the trajectory of light in terms of a general mathematical principle, which we now call *Fermat's principle of least time.* As he demonstrated, light would follow precisely the same bending trajectory determined by Snell if "light travels between two given points along the path of shortest time."

Heuristically this can be understood as follows. If light travels more quickly in the less dense medium, then to get from *A* to *B* (see figure) in the shortest time, it would make sense to travel a longer distance in this medium, and a shorter distance in the second medium in which it travels more slowly. Now, it cannot travel for too long in the first medium, otherwise the extra distance it travels would more than overcome the gain obtained by traveling at a faster speed. One path is just right, however, and this path turns out to involve a bending trajectory that exactly reproduces the trajectory Snell observed.

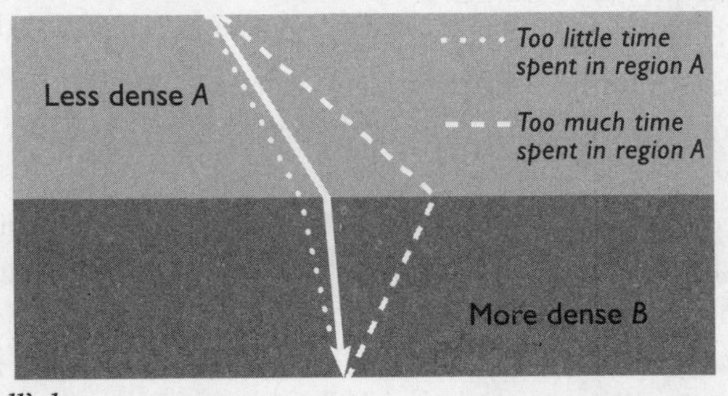

Snell's law

Fermat's principle of least time is a mathematically elegant way of determining the path light takes without recourse to any mechanistic description in terms of waves or particles. The only problem is that when one thinks about the physical basis of this result, it seems to suggest *intentionality*, so that, like a commuter in Monday-morning rush-hour listening to the traffic report, light somehow considers all possible paths before embarking on its voyage, and ultimately chooses the one that will get it to its destination fastest.

But the fascinating thing is that we don't need to ascribe any intentionality to light's wanderings. Fermat's principle is a wonderful example of an even more remarkable property of physics, a property that is central to the amazing and a priori unexpected fact that nature is comprehensible via mathematics. If there is any one property that was a guiding light for Richard Feynman's approach to physics, and essential to almost all of his discoveries, it was this one, which he thought was so important that he referred to it at least two different times during his Nobel Prize address. First, he wrote,

> It always seems odd to me that the fundamental laws of physics, when discovered, can appear in so many different forms that are not apparently identical at first, but, with a little mathematical fiddling you can show the relationship. . . . it was something I learned from experience. There is always another way to say the same thing that doesn't look at all like the way you said it before. . . . I think it is somehow a representation of the simplicity of nature. I don't know what it means, that nature chooses these curious forms, but maybe that is a way of defining simplicity. Perhaps a thing is simple if you can describe it fully in several different ways without immediately knowing that you are describing the same thing.

And later (and more important for what was to come), he added,

> Theories of the known, which are described by different physical ideas, may be equivalent in all their predictions and are hence scientifically indistinguishable. However, they are not psychologically identical when trying to move from that base into the unknown. For different views suggest different kinds of modifications which might be made and hence are not equivalent in the

hypotheses one generates from them in one's attempt to understand what is not yet understood.

Fermat's principle of least time clearly represents a striking example of this strange redundancy of physical law that so fascinated Feynman, and also of the differing "psychological utilities" of the different prescriptions. Thinking about the bending of light in terms of electric and magnetic forces at the interface between media reveals something about the properties of the media. Thinking about it in terms of the speed of light itself reveals something about light's intrinsic wavelike character. And thinking about it in terms of Fermat's principle may reveal nothing about specific forces or about the wave nature of light, but it illuminates something deep about the nature of motion. Happily, and importantly, all of these alternate descriptions result in identical predictions.

Thus we can rest easy. Light does not *know* it is taking the shortest path. It just *acts* like it does.

IT WASN'T THE principle of least time, however, but an even subtler idea that changed Feynman's life that fateful day in high school. As Feynman later described it, "When I was in high school, my physics teacher—whose name was Mr. Bader—called me down one day after physics class and said, 'You look bored; I want to tell you something interesting.' Then he told me something that I found absolutely fascinating, and have, since then, always found fascinating . . . the principle of least action." *Least action* may sound like an expression that is more appropriate to describing the behavior of a customer service representative at the phone company than a field like physics, which is, after all, centered around describing actions. But the least action principle is very similar to Fermat's principle of least time.

The principle of least time tells us that light always takes the path of shortest time. But what about baseballs and cannonballs, planets, and boomerangs? They don't necessarily behave so simply. Is there something other than *time* that is minimized whenever these objects follow the paths prescribed by the forces acting on them?

Consider any object in motion, say, a falling weight. Such an object is said to possess two different kinds of energy. One is *kinetic energy*, and it is related to the motion of objects (and derives from the Greek word for movement). The faster an object moves, the larger the kinetic energy. The other part of an object's energy is much subtler to ascertain, as reflected in its name: *potential energy*. This kind of energy may be hidden, but it is responsible for the ability of an object to do work later on. For example, a heavy weight falling off the top of a tall building will do more damage (and hence more work) smashing the roof of a car, than will a similar weight dropped from several inches above the car. Clearly the higher the object, the greater its potential to do work, and hence the greater its potential energy.

Now, what the least action principle states is that the *difference* between the kinetic energy of an object at any instant and its potential energy at the same instant, when calculated at each point along a path and then added up along the path, will be smaller for the actual path the object takes than for any other possible trajectory. An object somehow adjusts its motion so the kinetic energy and the potential energy are as closely matched, on average, as is possible.

If this seems mysterious and unintuitive, that is because it is mysterious and unintuitive. How on earth would anyone ever come up with this combination in the first place, much less apply it to the motion of everyday objects?

For this we thank the Italian mathematician-physicist Joseph Louis Lagrange, who is best known for his work on celestial mechanics. For example, he determined the points in the solar system where the gravitational attraction from the different planets precisely cancels the frame of reference of the orbiting body. They are called Lagrange points. NASA now sends numerous satellites out to these points so that they can remain in stable orbits and study the universe.

Lagrange's greatest contribution to physics, however, may have involved his reformulation of the laws of motion. Newton's laws relate the motion of objects to the net forces acting on them. However, Lagrange managed to show that Newton's laws of motion were precisely reproduced if one used the "action," which is the sum over a path of the differences between kinetic and potential energy, now appropriately called a Lagrangian, and then determined precisely what sorts of motion would produce those paths that minimized this quantity. The process of minimization, which required the use of calculus (also invented by Newton), gave very different mathematical descriptions of motion from Newton's laws, but, in the spirit of Feynman, they were mathematically identical, even if "psychologically" very different.

IT WAS THIS strange principle of least action, often called Lagrange's principle, that Mr. Bader introduced the teenaged Feynman to. Most teens would not have found it fascinating or even comprehensible, but Feynman did, or so he remembered when he was older.

However, if the young Feynman had any inkling at the time that this principle would return to completely color his own life story, he certainly didn't behave that way as he began to learn more about physics once he entered MIT. Quite the contrary. His best friend as an undergraduate at MIT, Ted Welton, with whom he worked through much of undergraduate and even graduate physics, later described Feynman's "maddening refusal to concede that Lagrange might have something useful to say about physics. The rest of us were appropriately impressed with the compactness, elegance, and utility of Lagrange's formulation, but Dick stubbornly insisted that real physics lay in identifying all the forces and properly resolving them into components."

Nature, like life, takes all sorts of strange twists and turns, and most important, it is largely insensitive to one's likes and dislikes. As much as Feynman tried early on to focus on understanding motion in a way that meshed with his naive intuition, his own trajectory to greatness involved a very different path. There was no unseen hand guiding him. Instead, he forced his intuition to bend to the demands of the problems of the time, rather than vice versa. The challenge required endless hours and days and months of hard work training his mind to wrap around a problem that the greatest minds in twentieth-century physics had, up to that point, not been able to solve.

When he really needed it, Feynman would find himself returning once again to the very principle that had turned him on to physics in the first place.

CHAPTER 2

The Quantum Universe

> I was always worried about the physics. If the idea looked lousy, I said it looked lousy. If it looked good, I said it looked good.
>
> —Richard Feynman

Feynman was fortunate to have stumbled upon Ted Welton in his sophomore year at MIT, while both were attending, as the only two sophomores, an advanced graduate course in theoretical physics. Kindred spirits, each had been checking advanced mathematics texts out of the library, and after a brief period of trying to outdo each other, they decided to collaborate "in the struggle against a crew of aggressive-looking seniors and graduate students" in the class.

Together they pushed each other to new heights, passing back and forth a notebook in which each would contribute solutions and questions on topics ranging from general relativity to quantum mechanics, each of which they apparently had taught themselves. Not only did this encourage Feynman's seemingly relentless quest to derive all of physics on his own terms, but also it provided some object lessons that would stay with him for the rest of his life. One in particular is worth noting. Feynman and Welton tried to determine the energy levels of electrons in a hydrogen atom by generalizing the standard equation of quantum mechanics, called the *Schrödinger equation*, to incorporate the results of Einstein's special relativity. In so doing they rediscovered what was actually a well-known equation, the Klein-Gordon equation. Unfortunately, after Welton urged Feynman to apply this equation to understand the hydrogen atom, the attempt produced results that completely disagreed with experimental results. This is not surprising because the Klein-Gordon equation was known to be the wrong equation to use to describe relativistic electrons, as the brilliant theoretical physicist Paul Dirac had demonstrated only a decade earlier, in the process of earning the Nobel Prize for deriving the right equation.

Feynman described his experience as a "terrible" but very important lesson that he never forgot. He learned not to rely on the beauty of a mathematical theory or its "marvelous formality," but rather to recognize that the test of a good theory was whether one could "bring it down against the real thing"—namely, experimental data.

Feynman and Welton were not learning all of physics completely on their own. They also attended classes. During the second semester of their sophomore year they had sufficiently impressed the professor of their theoretical physics course, Philip Morse, that he invited the two of them, along with another student, to study quantum mechanics with him in a private tutorial one afternoon a week during their junior year. Later he invited them to start a "real research" program in which they calculated properties of atoms more complicated than hydrogen, and in the process they also learned how to work the first generation of so-called calculating machines, another skill that would later serve Feynman well.

By the time of his final year as an undergraduate, Feynman had essentially mastered most of the undergraduate and graduate physics curricula, and he had already become excited enough by the prospect of a research career that he made the decision to proceed on to graduate school. In fact, his progress had been so impressive that during his junior year the physics department recommended that he be granted a bachelor's degree after three years instead of four. The university denied the recommendation, so instead, during his senior year, he continued his research and wrote a paper on the quantum mechanics of molecules that was published in the prestigious *Physical Review*, as was a paper on cosmic rays. He also took some time to reinforce his fundamental interest in the applications of physics, and enrolled in metallurgy and laboratory courses—courses that would later serve him well in Los Alamos—and even built an ingenious mechanism to measure the speeds of different rotating shafts.

Not everyone was convinced that Feynman should take the next major step in his education. Neither of his parents had completed a college education, and the rationale for their son completing yet another three or four years of study beyond an undergraduate degree was unclear. Richard's father, Melville Feynman, visited MIT in the fall of 1938 to speak to Professor Morse and ask if it was worth it, if his son was good enough. Morse answered that Feynman was the brightest undergraduate student he had ever encountered, and yes, graduate school not only was worth it, but was required if Feynman wanted to continue a career in science. The die was cast.

Feynman's preference was to stay on at MIT. However, wise physics professors generally encourage their students, even their best ones, to pursue their graduate studies at a new institution. It is important for students to get a broad exposure early in their career to the different styles of doing science, and to different focuses of interest, as spending an entire academic career at one institution can be limiting for many people. And so it was that Richard Feynman's senior dissertation advisor, John Slater, insisted that he go to graduate school elsewhere, telling him, "You should find out what the rest of the world is."

Feynman was offered a scholarship to Harvard for graduate school without even applying because he had won the William Lowell Putnam Mathematical Competition in 1939. This is the most prestigious and demanding national mathematics contest open to undergraduates, and was then in its second year. I remember when I was an undergraduate the very best mathematics students would join their university's team and solve practice problems for months ahead of the examination. No one solves all the problems on the exam, and in many years a significant fraction of the entrants fail to solve a single problem. The mathematics department at MIT had asked Feynman to join MIT's team for the competition in his senior year, and the gap between Feynman's score and the scores for all of the other entrants from across the country apparently astounded those grading the exam, so he was offered the Harvard prize scholarship. Feynman would later sometimes feign ignorance of formal mathematics when speaking about physics, but his Putnam score demonstrated that as a mathematician, he could compete with the very best in the world.

But Feynman turned down Harvard. He had decided he wanted to go to Princeton, I expect for the same reason that so many young physicists wanted to go there: that was where Einstein was. Princeton had accepted him and offered him a job as future Nobel laureate Eugene Wigner's research assistant. Fortunately for Feynman, he was assigned instead to a young assistant professor, John Archibald Wheeler, a man whose imagination matched Feynman's mathematical virtuosity.

In a remembrance of Feynman after his death, Wheeler recalled a discussion among the graduate admissions committee in the spring of 1939, during which one person raved about the fact that no one else applying to the university had math and physics aptitude scores anywhere near as high as Feynman's (he scored 100 percent in physics), while another member of the committee complained at the same time that they had never let anyone in with scores so low in history and English. Happily for the future of science, physics and math prevailed.

Interestingly, Wheeler did not describe another key issue, of which he may not have been aware: the so-called Jewish question. The head of the physics department at Princeton had written to Philip Morse about Feynman, asking about his religious affiliation, adding, "We have no definite rule against Jews but have to keep their proportion in our department reasonably small because of the difficulty of placing them." Ultimately it was decided that Feynman was not sufficiently Jewish "in manner" to get in the way. The fact that

Feynman, like many scientists, was essentially uninterested in religion never arose as part of the discussion.

MORE IMPORTANT THAN all of these external developments, however, was the fact that Feynman had now proceeded to the stage in his education where he could begin to think about the really exciting stuff—namely, the physics that didn't make sense. Science at the forefront is always on the verge of paradox and inconsistency, and like a bloodhound, great physicists focus precisely on these elements because that is where the true quarry lies.

The problem that Feynman later said he "fell in love with" as an undergraduate had been a familiar part of the centerpiece of theoretical physics for almost a century: the classical theory of electromagnetism. Like many deep problems, it can be simply stated. The force between two like charges is repulsive, and therefore it takes work to bring them closer together. The closer they get, the more work it takes. Now imagine a single electron. Think of it as a "ball" of charge with a certain radius. To bring all the charge together at this radius to make up the electron would thus take work. The energy built up by the work bringing the charge together is commonly called the *self-energy* of the electron.

The problem is that if we were to shrink the size of the electron down to a single point, the self-energy associated with the electron would go to infinity, because it takes an infinite amount of energy to bring all the charge together at a single point. This problem had been known for some time and various schemes had been put together to solve it, but the simplest was to assume that the electron really wasn't confined to a single point, but had a finite size.

By early in the twentieth century this issue took on a different perspective, however. With the development of quantum mechanics, the picture of electrons, and electric and magnetic fields, had completely changed. So-called wave-particle duality, for example, a part of quantum theory, said that both light *and* matter, in this case electrons, sometimes behaved as if they were particles and sometimes as if they were waves. As our understanding of the quantum universe grew, while the universe also got stranger and stranger, nevertheless some of the key puzzles of classical physics disappeared. But others remained, and the self-energy of the electron was one of them. In order to put this in context, we need to explore the quantum world a little bit.

Quantum mechanics has two central characteristics, both of which completely defy all of our standard intuition about the world. First, objects that are behaving quantum mechanically are the ultimate multitaskers. They are capable of being in many different configurations at the same time. This includes being in different places and doing different things simultaneously. For example, while an electron behaves almost like a spinning top, it can also act as if it is spinning around in many different directions at the same time.

If an electron acts as if it is spinning counterclockwise around an axis pointing up from the floor, we say it has *spin up*. If it is spinning clockwise, we say it has *spin down*. At any instant the probability that an electron has spin up may be 50 percent, and the probability that it has spin down may be 50 percent. If electrons behaved as our classical intuition would suggest, the implication would be that each electron we measure has either spin up or spin down, and that 50 percent of the electrons will be found to be in one configuration and 50 percent in the other.

In one sense this is true. If we measure electrons in this way, we will find that 50 percent are spin up and 50 percent are spin down. *But*, and this is a very important *but*, it is incorrect to assume that each electron is in one configuration or another before we make the measurement. In the language of quantum mechanics, each electron is in a "superposition of states of spin up and spin down" before the measurement. Put more succinctly, it is spinning both ways.

How do we know that the assumption that electrons are in one or another configuration is "incorrect"? It turns out that we can perform experiments whose results depend on what the electron is doing when we are not measuring it, and the results would come out differently if the electron had been behaving sensibly, that is, in one or another specific configuration between measurements.

The most famous example of this involves shooting electrons at a wall with two slits cut into it. Behind the wall is a scintillating screen, much like the screen on old-fashioned vacuum-tube televisions, that lights up wherever an electron hits it. If we don't measure the electrons between the time they leave the source and when they hit the screen, so that we cannot tell which slit each electron goes through, we would see a pattern of bright and dark patches emerge on the rear screen—precisely the kind of "interference pattern" that we would see for light or sound waves that traverse a two-slit device, or perhaps more familiarly, the pattern of alternating ripples and calm that often results when two streams of water converge together. Amazingly, this pattern emerges even if we send only a single electron toward the two slits at any time. The pattern thus suggests that somehow the electron "interferes" with itself after going through both slits at the same time.

At first glance this notion seems like nonsense, so we alter the experiment slightly. We put a nondestructive electron detector by each slit and then send the electrons through. Now we find that for each electron, one and only one detector will signal that an electron has gone through at any time, allowing us to determine that indeed each electron goes through one and only one slit, and moreover we can determine which slit each electron has gone through.

So far so good, but now comes the quantum kicker. If we examine the pattern on the screen after this seemingly innocent intervention, the new pattern is completely different from the old pattern. It now resembles the pattern we would get if we were shooting bullets at such a screen through the two-slit barrier—namely, there will be a bright spot behind each slit, and the rest will be dark.

So, like it or not, electrons and other quantum objects can perform classical magic by doing several different things at the same time, at least as long as we do not observe them in the process.

The other fundamental property at the heart of quantum mechanics involves the so-called *Heisenberg uncertainty principle*. What this principle says is that there are certain combinations of physical quantities, such as the position of a particle and its momentum (or speed), that we cannot measure at the same instant with absolute accuracy. No matter how good our microscope or measuring device is, multiplying the uncertainty in position by the uncertainty in momentum never results in zero; the product is always bigger than some number, and this number is called *Planck's constant.* It is this number that also determines the scale of the spacing between energy levels in atoms. In other words, if we measure the position very accurately so that the uncertainty in position is small, that means our knowledge of the momentum or speed of the particle must be very inaccurate, so that the product of the uncertainty in position and the uncertainty in momentum exceeds Planck's constant.

There are other such "Heisenberg pairs," like energy and time. If we measure the quantum mechanical state of a particle or an atom for a very short time, then there will be a big uncertainty in the measured energy of the particle or atom. In order to measure the energy accurately, we have to measure the object over a long time interval, in which case we cannot say precisely when the energy was being measured.

If this weren't bad enough, the quantum world gets even weirder once we add Einstein's theory of special relativity into the mix, in part because relativity puts mass and energy on the same footing. If we have enough energy available, we can create something with mass.

So, if we put all of these things together—quantum multiplexing, the Heisenberg uncertainty principle, and relativity—what do we get? We get a picture of electrons that is literally infinitely more confusing than the one presented by the classical theory, which already led to an infinite self-energy for the electron.

For example, whenever we try to picture an electron, it doesn't have to be just an electron! To understand this, let's return back to classical electromagnetism. One of the key features at the heart of this theory is the fact that if we shake an electron, it will emit elecromagnetic radiation, like light, or radio waves. This great discovery resulted from the groundbreaking nineteenth-century experiments of Michael Faraday, Hans Christian Oersted, and others, and the groundbreaking theoretical work of James Clerk Maxwell. Quantum mechanically, this observed phenomenon must still be predicted because if quantum mechanics is to properly describe the world, its predictions had better agree with observations. But the key new feature here is that quantum mechanics tells us to think of the radiation as being made up of individual *quanta*, or packets of energy, called photons.

Now let's return to the electron. The Heisenberg principle tells us that if we measure the electron for some finite time, there remains some finite uncertainty in knowing its exact energy. But if there is some uncertainty, how do we know we are measuring only the electron? For example, if the electron emits a photon carrying very little energy, the total energy of the system will change, albeit very slightly. But if we don't know the exact energy of the system, then we cannot say whether it has or hasn't emitted a low-energy photon. So what we are measuring really could be the energy of the electron plus a photon that it has emitted.

But why stop there? Perhaps the electron has emitted an infinite number of very-low-energy photons? If we watch the electron for long enough, we can both measure its energy very accurately and put a photon counter nearby to see if there are any photons around. In this case, what will have happened to all the photons that were traveling along with the electron in the interim? Simple: the electron can absorb all those photons before we get a chance to measure them.

The kind of photons that an electron can emit and reabsorb on a timescale so short that we cannot measure them are called *virtual particles*, and as I will describe later, Feynman recognized that when we include the effects of both relativity and quantum mechanics, there is no getting away from the existence of these particles. So when we think of an electron moving around, we now have to think of it as a pretty complicated object, with a cloud of virtual particles surrounding it.

Virtual particles play another important role in the quantum theory of electromagnetism. They change the way we think of electric and magnetic fields and the forces between particles. For example, say an electron emits a photon. This photon can then in turn interact with another particle, which can absorb it. Depending on the energy of the photon, this will result in a transfer of energy and momentum from one electron to another. But that is what we normally describe as the manifestation of the electromagnetic force between these two charged particles.

Indeed, as we will see, in the quantum world both electric and magnetic forces can be thought of as being caused by the exchange of virtual photons. Because the photon is massless, an emitted photon can carry an arbitrarily small amount of energy. Therefore, as the Heisenberg uncertainty principle tells us, the photon can travel an arbitrarily long distance (taking an arbitrarily long time) between particles before it must be reabsorbed in order that the energy it is carrying is returned back to the electron. It is precisely for this reason that the electromagnetic force between particles can act over long distances. If the photon had a mass, then it would always carry away a minimum energy, $E = mc^2$, where *m* is its mass, and in order for this violation of energy conservation to remain hidden within quantum uncertainties, the Heisenberg uncertainty principle implies that the photon must be reabsorbed by either the original electron or another electron within some fixed time, or equivalently within some fixed distance.

We are getting ahead of ourselves here, or at least ahead of Feynman at this time in his life, but introducing these complications at this point has a purpose. Because if all of this seems very complicated and hard to picture, join the crowd, especially the crowd in the era before World War II. This was the world of fundamental physics that Richard Feynman entered into as a student, and it was a world where the strange new rules seemed to produce nonsense. The classical infinite self-energy of the electron, for example, remained part of quantum theory, apparently owing to the fact that the electron could emit and reabsorb photons of arbitrarily high energy, as long as it did so over very short timescales.

But the confusion was even worse. The quantum theory fit well overall with experiment results. But whenever physicists tried to calculate predictions precisely to compare to accurate measurements—if they included the interchange of not just one photon between particles, for example, but more than one photon (a process that should happen more rarely than the exchange of a single photon) they found that the additional contribution due to this "higher order" effect was infinite. Moreover, the calculations in the quantum theory needed to explore these infinities were harrowingly difficult and tedious, taking the best minds at the time literally months to perform each such calculation.

While still an undergraduate, Feynman had an idea that

he carried with him to graduate school. What if the classical "picture" of electromagnetism, as I have described it, was wrong? What if, for example, there was a "new" rule that a charged particle could not interact with itself? That would, by fiat, get rid of the infinite self-energy of an electron because it could not interact with its own electric field. I emphasize that the infinity this new rule was designed to avoid is present in the pure classical theory, even without considering quantum mechanical effects.

But Feynman was even bolder. What if what we call the electromagnetic field, caused by an exchange of virtual photons between particles, also was a fiction? What if the whole of electromagnetism was due to a direct interaction between charged particles with no field present at all? Classically, electric and magnetic fields are completely determined by the motion of the charged particles producing them, so to Feynman the field was itself redundant. In other words, once the initial configuration of charges and their motion is specified, all of their subsequent motion could in principle be determined simply by considering the direct impact of the charges on one another.

Moreover, Feynman reasoned that if we could dispense with the electromagnetic field in the classical theory, this might solve the quantum problems as well, because if we could dispense with all of the infinite number of photons running around the calculations in the quantum theory and just deal with charged particles, perhaps we could get sensible answers. As he put it in his Nobel address, "Well, it seemed to me quite evident that the idea that a particle acts on itself is not a necessary one—it is a sort of silly one, as a matter of fact. And so I suggested to myself that electrons cannot act on themselves; they can only act on other electrons. That means there is no field at all. There was a direct interaction between charges, albeit with a delay."

These were bold ideas, and Feynman brought them to graduate school at Princeton, and to John Archibald Wheeler, who was precisely the man to bounce them off of. I knew John Wheeler as a most gentle and cordial soul, polite and considerate to a fault, like a perfect southern gentleman (even though he was from Ohio). But when he talked about physics, he suddenly became bold and fearless. In the words of one of his Princeton colleagues at the time, "Somewhere among those polite facades there was a tiger loose . . . who had the courage to look at any crazy problem." This kind of fearlessness matched Feynman's intellectual predilections exactly. I remember causing ripples of laughter when I quoted Feynman once as saying in a letter to a potential young physicist, "Damn the torpedoes. Full speed ahead." Feynman of course was aping Admiral David Farragut, but that historical fact seemed irrelevant. That phrase applied equally well to both Feynman and Wheeler.

It was a match made in heaven. What followed at Princeton was an intense three-year period of intellectual give-and-take between the two resonant minds—physics as it should be done. Neither man would immediately discount the crazy ideas of the other. As Wheeler later wrote, "I am eternally grateful for the fortune that brought us together on more than one fascinating enterprise. . . . Discussions turned into laughter, laughter into jokes, and jokes into more to-and-fro and more ideas. . . . From more than one of my courses he knew my faith that whatever is important is at bottom utterly simple."

When Feynman first brought his crazy idea to Wheeler, it was not met with derision. Instead, Wheeler immediately pointed out its flaws, reinforcing the axiom "Fortune Favors the Prepared Mind," for Wheeler too had been thinking along very similar lines.

Feynman had realized earlier one glaring fault with his idea. It is well known that it takes more work to accelerate a charged particle than a neutral one, because in the process of acceleration a charged particle emits radiation and dissipates energy. Thus a charged particle does seem to act on itself by producing an extra resistance (called *radiation resistance*) to being pushed around. Feynman had hoped that somehow he could resolve this problem by considering the reaction back on the particle, not by itself, but by the induced motion of all of the other charges in nature that would be affected by their interactions with the first particle. Namely, the force from the first particle on the other particles would cause them to move, and their motion would produce electric currents that could then react back on the first particle.

When he first heard about these ideas, Wheeler responded by pointing out that if this were the case, the radiation resistance produced by the first particle would depend on the location of these other charges, which it doesn't, and moreover would be delayed because no signal could travel faster than the speed of light. It would hence take time for the first particle to interact with the second (some distance away) and even more time for the second particle to then interact back with the first particle—resulting in a back reaction that would be considerably delayed in time compared to the initial motion of the first particle.

But then Wheeler suggested an even crazier idea: what if the return action by these other charges somehow acted backward in time? Then instead of the back reaction of these particles on the first particle occurring well after the first particle had started to move, it might occur at the exact same time the first particle started to move. At this point a sensible novice might say, "Hold on there, isn't that crazy? If particles can react backward in time, then doesn't this violate sacred principles of physics like causality, which requires causes to happen before effects?"

But while allowing for backward back-reaction opens up such a possibility in principle, to find out if it really causes problems, physicists must be more precise and actually perform the calculations first. And this is what Feynman and Wheeler did. They were playing around to see if they could fix their problems without creating new ones, and they were willing to suspend disbelief until their results required them not to.

First off, based on his prior thinking about these issues, Wheeler was able to work out with Feynman almost immediately that in this case the radiation reaction could be derived to be independent of the location of the other charges, and could also in principle be made to occur at the appropriate time, and not at some later, delayed, time.

Wheeler's proposal had its own problems, but it got Feynman thinking, and calculating. He worked through the details and determined precisely how much of the backward-in-time reaction between particles was needed to make things work out just right, and as was typical of Feynman, he then also checked a lot of different examples to make sure that this idea would not produce crazy phenomena that are not observed, or violations of common sense. He challenged his friends to find an example that might stump him, and he showed that as long as in every direction in the universe there was 100 percent certainty that one would ultimately encounter a charged particle that could interact back with the original particle, one could never use these crazy backward-in-time interactions to produce a device that could turn on before the on button is pushed, or anything like that.

As Humphrey Bogart might have said, it was the beginning of a beautiful friendship. Whereas Feynman had mathematical brilliance and startlingly good insight, Wheeler had experience and perspective. Wheeler was able to quickly shoot down some of Feynman's misconceptions and suggest improvements, but he had an open mind and encouraged Feynman to explore and to gain calculational experience that was adequate to match his talents. Once Feynman combined the two, he would be almost unstoppable.

CHAPTER 3

A New Way of Thinking

> An idea which looks completely paradoxical at first, if analyzed to completion in all its details and in experimental situations, may in fact not be paradoxical.
>
> —RICHARD FEYNMAN

Despite the assurances of Richard's undergraduate professors, Melville Feynman did not lay his concerns about his son's future to rest. After Richard had begun his working relationship with John Archibald Wheeler in graduate school, Melville made the trek to Princeton to check once more on his progress and prospects. Once again, he was told that Richard had a brilliant future ahead of him, independent of his "simple background" or possible "anti-Jewish prejudice," as Melville phrased things. Wheeler may have been sugarcoating reality, or merely reflecting his own ecumenical bent. While still a student he had been the founder and president of the Federation of Church and Synagogue Youth.

Nevertheless, even any lingering anti-Semitism in academia would not have been sufficient to halt Richard Feynman's march forward. He was simply too good and having too much fun. Only a fool would not recognize his genius and his potential. As they would continue to throughout his life, Feynman's fascination with physics, and his ability to solve problems others couldn't, stretched across the spectrum of the physical world, from the esoteric to the seemingly mundane.

Everywhere there were glimpses of his playful intensity. The Wheeler children used to love his visits, when he would often amuse them with tricks. Wheeler remembered one afternoon when Feynman asked for a tin can and told the children that he could tell whether solid or liquid was inside without even opening it or looking at the label. "How?" came a chorus of young voices. "By the way it turns when I toss it up in the air," he answered, and sure enough, he was right.

Feynman's own childlike excitement about the world meant that his popularity with children remained unabated, as reflected in a letter written in 1947 by the physicist Freeman Dyson, who was a graduate student at Cornell when Feynman was an assistant professor there. Describing a party at the home of the physicist Hans Bethe in honor of a distinguished visitor, Dyson remembered that Bethe's five-year-old son, Henry, kept complaining that Feynman was not there, saying, "I want Dick. You told me Dick was coming." Ultimately Feynman arrived, dashed upstairs, and then proceeded to play noisily with Henry, stopping all conversation down below.

Even as Feynman entertained Wheeler's children, he and Wheeler continued to amuse each other as they worked throughout the year to explore their exotic ideas on ridding classical electromagnetism of the problem of infinite self-interaction of charged particles, via strange backward-in-time interactions with external absorbers located out in an infinite universe.

Feynman's motivation for continuing this work was straightforward. He wanted to solve a mathematical problem in classical electromagnetism with the hope of ultimately addressing the more serious problems that arose in the quantum theory. Wheeler, on the other hand, had an even crazier notion he wanted to develop to explain the new particles that were being observed in cosmic rays and ultimately in nuclear physics experiments: maybe all elementary particles were just made of different combinations of electrons, somehow interacting differently with the outside world. The notion was crazy, but at least it helped maintain his own enthusiasm for the work they were doing.

Feynman's own playful attitude toward the inevitable frustrations and stumbling blocks associated with theoretical work in physics is exemplified by one of the earliest letters he wrote his mother, shortly after starting graduate school and before his work with Wheeler had moved in the direction of reexamining electromagnetism:

> Last week things were going fast and neat as all heck, but now I'm hitting some mathematical difficulties which I will either surmount, walk around, or go a different way—all of which consumes all my time—but I like to do very much and am very happy indeed. I have never thought so much so steadily about one problem—so if I get nowhere I really will be very disturbed—However, I have already gotten somewhere, quite far—and to Prof. Wheeler's satisfaction. However, the problem is not at

completion although I'm just beginning to see how far it is to the end and how we might get there (although aforementioned mathematical difficulties loom ahead)—SOME FUN!

Feynman's idea of fun included prevailing over mathematical difficulties—one of the many attributes that probably separated him from the man on the street.

After an intense few months of give-and-take with Wheeler in the fall and winter of 1940–41 working on their new ideas for electromagnetism, Wheeler finally gave Feynman a chance to present these ideas, not to graduate students, but to professional physicists, through the Princeton physics department seminar. But this was not to be just any group of colleagues. Eugene Wigner, himself a later Nobel Prize winner, ran the seminar and invited, among others, a special cast of characters: the famous mathematician John von Neumann; the formidable Nobel Prize winner and one of the developers of quantum mechanics, Wolfgang Pauli, who was visiting from Zurich; and none other than Albert Einstein, who had expressed an interest in attending (perhaps egged on by contact with Wheeler).

I have tried to imagine myself in Feynman's place, as a graduate student speaking among such a group. This would not be an easy crowd to please, independent of their eminence. Pauli, for example, was known to jump up and take the chalk out of the hands of speakers with whom he disagreed.

Feynman nevertheless prepared his talk and once he began, the physics took over and any residual nervousness disappeared. As expected, Pauli objected, concerned about whether the use of the backward-in-time reactions might have implied that one was simply working backward mathematically from the correct answer and not actually deriving anything new. He was also concerned about the "action-at-a-distance" aspect of the ideas, once one had dispensed with the fields that usually transport the forces and information, and he asked Einstein whether this might be incompatible with his own work on general relativity. Amusingly, Einstein humbly responded that there might be a conflict, but after all his own theory of gravitation (which the rest of the physics community has regarded as the most significant single piece of work since Newton) was "not so well established." Actually, Einstein was sympathetic to the notion of using backward-in-time as well as forward-in-time solutions, as Wheeler later recalled, when he and Feynman went to visit Einstein at his Mercer Street home to talk further about their work.

The problem is that one of the most obvious features of the physical world, manifest from the moment we wake up each day, is that the future is different from the past. This is true not only for human experience, but also for the behavior of inanimate objects. When we put milk in our coffee and stir it, we will never see that milk at some point in the future coalesce into separate droplets like it appeared when we first poured it into the coffee. The question is: does this apparent temporal irreversibility in nature arise because of an asymmetry in microscopic processes, or is it only appropriate for the macroscopic world we experience?

Like Feynman and Wheeler, Einstein believed that the microscopic equations of physics should be independent of the arrow of time—namely, the apparent irreversibility of phenomena in the macroscopic world arises because certain configurations are far more likely to arise naturally when many particles are involved than are other configurations. In the case of Feynman and Wheeler's ideas, as Feynman had shown to his fellow graduate student, physics behaved sensibly in the bulk—that is, the future was different from the past in spite of the weird backward-in-time interaction he and Wheeler included. This was precisely because the probabilities associated with the behavior of the rest of the presumably infinite number of other charges in the universe responding to the motion of the charge in question produced the kind of macroscopic irreversibility we are used to seeing in the world around us.

Later, in 1965, physicists discovered, much to their surprise, that certain microscopic processes for elementary particles do have an arrow of time associated with them—namely, the rates for a process and its time-reversed version are slightly different. This result was so surprising that it garnered the Nobel Prize for the experimentalists involved. Nevertheless, while this effect may play an important role in understanding certain features of our universe, including perhaps why we live in a universe of matter and not antimatter, conventional wisdom still suggests that the macroscopic arrow of time is associated with the tendency for disorder to increase, and arises, not from microscopic physics, but from macroscopic probabilities, as Einstein, Feynman, and Wheeler had presumed.

ULTIMATELY ALL THIS Sturm und Drang associated with the interpretation of Feynman and Wheeler's ideas was misplaced. The theoretical ideas they had proposed ended up being more or less wrong, in that their proposals didn't ultimately correspond to reality. Electrons do have self-interactions, and electromagnetic fields, including those involving virtual particles, are real. Feynman summed it up well a decade later when he wrote to Wheeler, "So I think we guessed wrong in 1941. Do you agree?" History has recorded no response from Wheeler, but by then the evidence was indisputable.

So what was the point of all of this work? Well, in science almost every significant new idea is wrong, either trivially wrong (there is a mathematical error) or more substantially wrong (as beautiful as the idea is, nature chooses not to exploit it). If that were not the case, then pushing the frontiers of science forward would be almost too easy.

In light of this, scientists have two choices. Either they can choose to follow well-trodden ground and push solid results a tad further with a reasonable assurance of success. Or they can strike out into new and dangerous territory, where there are no guarantees and they have to be prepared for failure. This might seem depressing, but in the process of exploring all the dead ends and blind alleys, scientists build up experience and intuition and a set of useful tools. Beyond this, unexpected ideas resulting from proposals that lead nowhere, at least as far as the original problem is concerned, nevertheless sometimes carry scientists in a direction that was completely unanticipated, and which every now and then can hold the key to progress. Sometimes ideas that don't work in one area of science end up being just what was needed to break a logjam elsewhere. As we will see, so it was with Richard Feynman's long journey through the wilderness of electrodynamics.

... ...

Amid the turbulent intellectual flow in Feynman's life during this period, his personal life evolved deeply as well. Ever since he had been a young man, a child almost, he had known, admired, and dreamed about a certain girl, a girl who possessed qualities that weren't manifest in him: artistic and musical talents, and the social confidence and grace that often accompany them. Arline Greenbaum had become a presence in his life early on in high school. He had met her at a party when he was fifteen and she was thirteen. She must have had everything he was looking for. She played the piano, danced, and painted. By the time he had entered MIT, she had become a fixture in his family life, painting a parrot on the door of a clothes closet for the family, and teaching piano to his sister Joan and then taking her for walks afterward.

We will never know if these kindnesses were Arline's way of ingratiating herself with Richard, but it was clear that she had decided he was the young man for her, and he too was smitten. Joan later claimed that by the time Richard entered MIT, when he was seventeen, the rest of the family knew that one day they would be married. They were right. Arline had visited him at his fraternity in Boston on weekends during his early years at college, and by the time he was a junior, he had proposed, and she had accepted.

Richard and Arline were soul mates. They were not clones of each other, but symbiotic opposites—each completed the other. Arline admired Richard's obvious scientific brilliance, and Richard clearly adored the fact that she loved and understood things he could barely appreciate at that time. But what they shared, most important of all, was a love of life and a spirit of adventure.

Arline makes her way into this scientific biography at this point not merely because she was Feynman's first, and perhaps deepest, love, but because her spirit provided him with the vital encouragement he needed to keep going, to find new roads, to break traditions, scientific and otherwise.

Their correspondence during the five years between the time of his proposal and her death from tuberculosis is remarkably touching and moving. Filled with naive hope, combined with mutual love and respect, they reflect two young people who were determined to make their own way in the world no matter what the obstacles.

In June of 1941, when Richard was well along in his graduate career, and a year before their marriage, Arline wrote to fill him in on her visits to doctors (it took many misdiagnoses before they eventually got her condition correct), but the letter focuses on him, not her:

> Richard sweetheart I love you. . . . we still have a little more to learn in this game of life and chess—and I don't want to have you sacrifice anything for me. . . . I know you must be working very hard trying to get your paper out—and do other problems on the side—I'm awfully happy tho' that you're going to publish something—it gives me a very special thrill when your work is acknowledged for its value—I want you to continue and really give the world and science all you can . . . and if you receive criticisms—remember everyone loves differently.

Arline knew Richard as no one else did, and in so doing she had the power both to embarrass him and to drive him forward to hold true to his beliefs. Most important among these were honesty and the courage to make his own choices. The title of one of his famous autobiographical books, "*What Do* You *Care What Other People Think?*" is the question she often repeated when she caught him in a timid or insecure moment, such as when she sent him a box of pencils, each engraved with the phrase "Richard darling, I love you! Putsie" (Putsie was his pet name for her), and caught him slicing off the words in case Professor Wheeler might see them when they were working together. If Feynman had the courage of his convictions, and ultimately the courage to go his own way in the world, both intellectually and otherwise, it was in no small part due to Arline and his memory of her.

While Melville Feynman was concerned about his son's career direction, Richard's mother, Lucille, was equally concerned about his personal one. She surely loved Arline, but she wrote to him, late in his graduate career, like many a Jewish mother, concerned that Arline would be a drag on his ability to work and gain a job and on his finances. Arline's illness would require special care, and time and money, and Lucille was worried that Richard didn't have enough of any of these things.

Richard responded, within weeks of receiving his PhD and marrying Arline, in June of 1942, remarkably dispassionately:

> I'm not dopey enough to tie up my whole life in the future because of some promise I made in the past—

under different circumstances. . . . I want to marry Arline because I love her—which means I want to take care of her. That is all there is to it. . . .

I have, however, other desires and aims in the world. One of them is to contribute as much to physics as I can. This is, in my mind, of even more importance than my love for Arline.

It is therefore especially fortunate that, as I can see (guess) my getting married will interfere very slightly, if at all with my main job in life. I am quite sure I can do both at once. (There is even the possibility that the consequent happiness of being married—and the constant encouragement and sympathy of my wife will aid in my endeavor—but actually in the past my love hasn't affected my physics much, and I don't really suppose it will be too great an assistance in the future.)

Since I feel I can carry on my main job, and still enjoy the luxury of taking care of someone I love—I intend to be married shortly.

Whether or not his love affected his physics, Arline had clearly reinforced his determination to follow his ideas wherever they might lead. She had helped ensure his intellectual integrity, and if the words in his letter seem somewhat cold and dispassionate, Arline might have been encouraged by them had she ever read them, because they reflected the kind of rational thinking she so wanted to foster in the man she so loved and admired.

She might have been equally moved by a heart-wrenching event that happened much later, on the dark day of her death, June 16, 1945, six weeks before the atomic bomb

Richard had worked to build was exploded over Hiroshima. After she breathed her last breath in the hospital room, he kissed her, and the nurse recorded the time of death as 9:21 p.m. He later discovered that the clock by her bedside had stopped at precisely the same time. A less rational mind might have found this cause for spiritual wonder or enlightenment—the kind of phenomena that makes people believe in a higher cosmic intelligence. But Feynman knew the clock was fragile. He had fixed it several times and he reasoned that the nurse must have picked it up and disturbed it to check the time of Arline's death. He would display the same kind of intellectual focus and determination to continue down a road he began in 1941, one that would ultimately, profoundly, and irrevocably change the way we think about the world.

THE WRITER LOUISE Bogan once said, "The initial mystery that attends any journey is: how did the traveler reach his starting point in the first place?" For Feynman's journey, like many epic voyages, the beginning was simple enough. He and Wheeler had completed their work demonstrating that classical electromagnetism could be cast in a form that involved only direct interactions, albeit forward and backward in time, between different charged particles. In so doing, one could obviate the problem of the infinite self-energy of any individual charged particle. The next challenge was to see if this theory could be brought into accord with quantum mechanics, and possibly resolve the thornier mathematical problems that resulted in a quantum theory of electromagnetism.

The only problem was that their rather exotic theory—

which was rigged with interactions at different times and places in order to be equivalent in its predictions with the results of classical electromagnetism, and which had electric and magnetic fields that transmitted these interactions—required a mathematical form that quantum mechanics couldn't handle at the time. The problem originated because of the interactions between particles at different times, or as Feynman later put it, "The path of one particle at a given time is affected by the path of another at a different time. If you try to describe, therefore, things . . . telling what the present conditions of the particles are, and how these present conditions will affect the future—you see it is impossible with particles alone, because something the particles did in the past is going to affect the future." Up to that point quantum mechanics was based on a simple principle. If we somehow knew or were told the quantum state of a system at one time, the equations of quantum mechanics allowed us to determine precisely the subsequent dynamical evolution of the system. Of course, knowing precisely the dynamical evolution of the system is not the same as predicting exactly what we would subsequently measure. The dynamical evolution of a quantum system involves determining exactly, not the final state of the system, but rather a set of probabilities that tells us what the likelihood is the system being measured will be in some specific state at a later time.

The problem is that electrodynamics as formulated by Feynman and Wheeler required knowing the positions and motion of many other particles at many different times in order to determine the state of any one given particle at any given time. In such a case, the standard quantum methods for determining the subsequent dynamical evolution of this particle failed.

Feynman had succeeded during the fall and winter of 1941–42 in formulating their theory in a host of different, if mathematically equivalent, ways. During the process he had discovered that he could rewrite the theory completely in terms of the very principle that he had so renounced while an undergraduate.

Remember that Feynman had learned in high school that there was a formulation of the laws of motion which was based not on what was happening at a single time, but what happened at all times: the formalism of Lagrange and his principle of least action.

Recall also that the least action principle tells us that in order to determine the actual classical trajectory of a particle, we can consider all possible paths of the particle between its beginning and end points and then determine which one has the smallest average value for the action (defined as the differences between two different parts of the total energy of the particle—the so-called kinetic and potential energies—appropriately summed over each path). This was the principle that was too elegant for Feynman, who preferred to calculate trajectories by considering forces at every point and using Newton's laws. The idea that he had to worry about the entire path of a particle in order to calculate its behavior at any point seemed unphysical to Feynman at the time.

But Feynman the graduate student had discovered that his theory with Wheeler could be recast completely in terms of an action principle, described purely by the trajectories of charged particles over time, with no need to consider electric and magnetic fields. In retrospect it seems clear why such formalism, which focused on the paths of particles, was appropriate to describe the Feynman-Wheeler theory. After all, such paths are essentially what defined their theory, which depended completely on the interactions of particles moving along different trajectories in time. Therefore, to build a quantum theory, Feynman decided he would need to figure out how to do quantum mechanics for a system like the one he and Wheeler were considering, whose classical dynamics could be determined by such an action principle, but *not* by more conventional methods.

Physics, or at least the physics that Feynman and Wheeler were imagining, had driven Feynman to a place he never would have expected to be a half-dozen years earlier. The transformation in his thinking following his intensive efforts to explore their new theory had been dramatic. He was now convinced that focusing on events at a fixed time was not the way to think, and that the action principle, based on exploring complete trajectories through space and time, was. As he later wrote, "We have in [the action principle] a thing that describes the character of the path throughout all of space and time. The behavior of nature is determined by saying her whole-space-time path has a certain character." But how could this principle be translated into quantum mechanics, which thus far depended so crucially on defining a system at one time in order to calculate what would happen at later times? For Feynman, the key to the answer came from a fortuitous beer party in Princeton. But to appreciate this key, we first have to make a short detour to revisit our picture of the mysterious quantum world that Feynman was about to change.

CHAPTER 4

Alice in Quantumland

> The Universe is not only queerer than we suppose, but queerer than we *can* suppose.
>
> J. B. S. HALDANE, 1924

While the distinguished British scientist J. B. S. Haldane was a biologist and not a physicist, his statement about the universe could not be more apt, at least to the realm of quantum mechanics that Richard Feynman was about to conquer. For, as we have seen, at the small scales where quantum mechanical effects become significant, particles can appear to be in many different places at once, while also doing many different things at the same time in each place.

The mathematical quantity that can account for all of this apparent lunacy is the function discovered by the famous Austrian physicist Erwin Schrödinger, who derived what became the conventional understanding of quantum mechanics during a busy two-week period in which he too was doing many different things at the same time—in the midst of trysts with perhaps two different women while holed up in a mountain chalet. It probably was the perfect atmosphere to imagine a world where all of the classical rules of behavior would ultimately be broken.

This function of Schrödinger's is called the *wave function* of an object, and it accounts for the mysterious fact, at the heart of quantum mechanics, that all particles behave in some sense like waves, and all waves behave in some sense like particles—the difference between a particle and a wave being that a particle is located at a specific point, whereas a wave is spread out over some region.

So, if a particle, which isn't spread out, is to be described by something that behaves like a wave and is spread out, the wave function must accommodate this fact. As Max Born later demonstrated, this was possible if the wave function, which itself might behave like a wave, did not describe the particle itself but rather the *probability* of finding the particle at any given place in space at a specific time. If the wave function, and hence the probability of finding a particle, is nonzero at many different places, then the particle acts like it is in many different places at any one time.

So far so good, even if the notion itself seems crazy. But there is one more crucial bit of craziness at the heart of quantum mechanics, and I should stress that physicists do not have a fundamental understanding of why nature behaves this way, except to say that it does. If the laws of quantum physics determine the behavior of the wave function, then physics tells us that given the wave function of a particle at one time, quantum mechanics in principle allows us to calculate, in a completely deterministic way, the wave function of the particle at a later time. Up to this point it is just like Newton's laws, which tell us how the classical motion of a baseball evolves in time, or Maxwell's equations, which tell us how electromagnetic waves evolve in time. The difference is that in quantum mechanics the quantity that evolves in time in a deterministic manner is not directly observable, but rather is a set of probabilities for making certain observations, in this case for determining the particle to be at a certain place at a certain time.

This is strange enough, but it further turns out that the wave function itself does not directly describe the probability of finding a particle at a given place at some time. Instead it is the *square* of the wave function that gives the probabilities. This one fact is responsible for all of the strangeness of the quantum mechanical world because it explains why particles can behave precisely as waves, as I will describe now.

First, note that the probabilities of things we measure must be positive (we would never say that there is a probability of minus 1 percent of finding something) and the square of a quantity is also always positive, so quantum mechanics predicts positive probabilities—which is a good thing. But it also implies that the wave function itself can be either positive or negative, since, say, $-\frac{1}{2}$ and $+\frac{1}{2}$ both yield the same number ($+\frac{1}{4}$) when squared.

If it were the wave function that described the probability of finding some particle at some location x, then if I had two identical particles, the probability of finding either particle at location x would be the sum of the two individual (and each necessarily positive) wave functions. However, because the square of the wave function is what determines the probability of finding particles, and because the square of the sum of two numbers is *not* equal to the sum of the individual squares, things can get much more interesting in quantum mechanics.

Let's say the value of the wave function that corresponds to finding particle *A* at position *x* is P1, and the value of the wave function that corresponds to finding particle *B* at position *x* is P2, then quantum mechanics tells us that the probability of finding either particle *A* or *B* at position *x* is now $(P1 + P2)^2$. Let's say P1 = ½ and P2 = –½. Then if we only had one particle, say particle *A*, the probability of finding it at position *x* would be $(½)^2$ = ¼. Similarly the probability of finding particle *B* at position *x* would be $(–½)^2$ = ¼. However, if there are two particles, the probability of finding *either* particle at position *x* is $((½) + (–½))^2 = 0$.

This phenomenon, which on the surface seems ridiculous, is in fact familiar for waves, say, sound waves. Such waves can *interfere* with each other so that, for example, waves on a string can interfere and produce locations on the string, called *nodes*, that do not move at all. Similarly, if sound waves are coming from two different speakers in a room, we might find, if we were to walk around the room, certain locations where the waves cancel each other out, or, as physicists say, *negatively interfere* with each other. (Acoustic experts design concert halls so that hopefully there are no such "dead spots.")

What quantum mechanics, with probabilities being determined by the square of the wave function, tells us is that particles too can *interfere* with each other, so that if there are two particles in a box, the probability of finding either of them at a given location can end up being less than the probability of finding one where only a single particle is in the box.

When waves interfere, it is the height, or *amplitude*, of the resulting wave that is affected by this interference, and the amplitude can be positive or negative depending on whether one is at a peak or a trough in the wave. So another name for the wave function of a particle is its *probability amplitude*, which can be positive or negative.

And just as for regular amplitudes for sound waves, separate probability amplitudes for different particles can cancel each other out.

It is precisely this mathematics that is behind the behavior of electrons shot at a scintillating screen, as described in chapter 2. Here we find that an electron can actually interfere with itself because electrons have a nonzero probability of being in many different places at any one time.

Let's first think about how to calculate probabilities in a sensible, classical world. Consider choosing to travel from town *a* to town *c* by taking some specific route through town *b*. Let the probability of choosing some route from *a* to *b* be given by P(*ab*), and then the probability of choosing some specific route from *b* to *c* be P(*bc*). Then, if we assume that what happens at *b* is completely independent of what happens at *a* and *c*, the probability of traveling from *a* to *c* along a specific route that goes through town *b* is simply given by the product of the two probabilities, $P(abc) = P(ab) \times P(bc)$. For example, say there is a 50 percent chance of taking some route from *a* to *b*, and then a 50 percent chance of taking some route from *b* to *c*. Then if we were to send four cars out, two will make it to *b* on the chosen route, and of those two, one will take the next chosen route from *b* to *c*. Thus there is a 25 percent ($.5 \times .5$) probability of taking the required route all the way from beginning to end.

Now, say we don't care which particular point *b* is visited between *a* and *c*. Then the probability of traveling from *a* to *c*, given by P(*ac*), is simply the sum of the probabilities P(*abc*) of choosing to go through *any* point *b* between *a* to *c*.

The reason this makes sense is that classically if we are going from *a* to *c*, and *b* represents the totality of different towns we can cross through, say, halfway from both *a* and *c*, then we have to go through one of them during our journey (see figure).

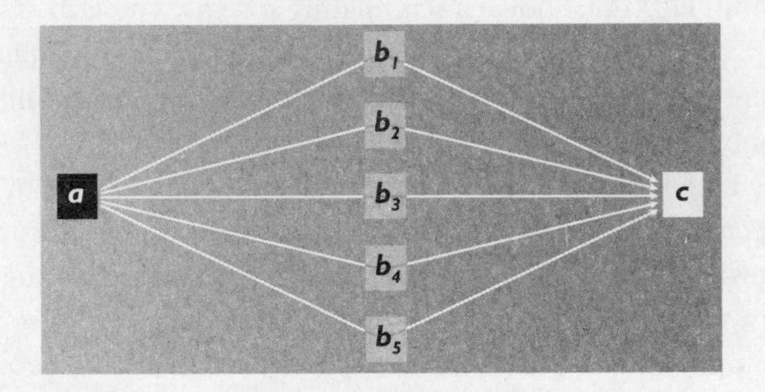

(Since this picture is reminiscent of the earlier pictures of light rays, then we could say that if the example in question involved light rays going from *a* to *c*, then we could use the principle of least time to determine that the probability of going through one of the routes, that of least time, was 100 percent, and the probability of taking any other route was zero.)

The problem is that things don't work this way in quantum mechanics. Because probabilities are determined by the squares of probability amplitudes to go from one place to another, the probability to go from *a* to *c* is *not* given by the sum of the probabilities to go from *a* to *c* via any definite intermediate point *b*. This is because in quantum mechanics it is the separate probability *amplitudes* for each part of the route that multiply and not the probabilities themselves. Thus, the probability amplitude to go from *a* to *c* through some definite point *b* is given by multiplying the probability amplitude to go from *a* to *b* times the probability amplitude to go from *b* to *c*.

If we don't specify which point *b* to travel through, the probability amplitude to go from *a* to *c* is again given by the sum of the product of probability amplitudes to go from *a* to *b* and from *b* to *c*, for all possible *b*'s. But this means that actual probability is now given as the *square* of the sum of these products. Since some terms in the sum can be negative, the crazy quantum behavior I discussed in chapter 2 for electrons hitting a screen can occur. Namely, if we don't measure which of two points, say *b* and *b'*, a particle traverses as it travels through one of two slits between *a* and *c*, then the probability of arriving at point *c* on the screen is determined by the sum of the squares of the probability amplitudes for the two different allowed paths. If we do measure which point, *b* or *b'*, the particle traverses in between *a* and *c*, then the probability is simply the square of the probability amplitude for a single path. In the case of many electrons shot one at a time, the final pattern on the screen in the former case will be determined by adding the squares of the sum of probability amplitudes for each of the two possible paths for each particle, while in the latter case it will be determined by adding the squares of the probability amplitudes for each path separately taken by each electron. Again, because the square of a sum of numbers is different from the sum of squares of these numbers, the former probability can differ dramatically from the latter. And as we have seen, if the particles are electrons, the results are indeed different if we don't measure the particle between beginning and end points compared to what happens if we do.

Quantum mechanics works, whether or not it makes sense.

IT IS PRECISELY this seemingly nonsensical aspect of quantum mechanics that Richard Feynman focused on. As he later put it, the classical picture is wrong if the statement that the position of the particle midway on its trip from *a* to *c* actually takes some specific value, *b*, is wrong. Quantum mechanics instead allows for all possible *paths*, with all values of *b* to be chosen at the same time.

The question that Feynman then asked is, Can quantum mechanics be framed in terms of the *paths* associated with probability amplitudes rather than the probability amplitudes themselves? It turned out that he was not the first one who had asked this question, though he was the first to derive the answer.

CHAPTER 5

Endings and Beginnings

> Instead of putting the thing into the mind, or psychology, I put it into a number.
>
> —Richard Feynman

As he was struggling to come up with a way to formulate quantum mechanics to accommodate his strange theory with Wheeler, Richard Feynman attended what he later called a "beer party" at the Nassau tavern in Princeton. There he met the European physicist Herbert Jehle, who was visiting at the time and asked Richard what he was working on. Feynman said he was trying to come up with a way to develop quantum mechanics around an action principle. Jehle told him about a paper by one of the fathers of quantum mechanics, the remarkable physicist Paul Dirac, that might just hold the key. As he remembered it, Dirac had proposed how one might use the quantity from which the action is calculated (which, recall from chapter 1, is the Lagrangian—equal to the difference in kinetic and potential energies of a system of particles) in the context of quantum mechanics.

The very next day they went to the Princeton library to go through Dirac's 1933 paper, suitably titled "The Lagrangian in Quantum Mechanics." In that paper Dirac brilliantly and presciently suggested that "there are reasons for believing that the Lagrangian [approach] is more fundamental" than other approaches because (a) it is related to the action principle, and (b) (of vital importance for Feynman's later work, but a fact he wasn't thinking about at the time) the Lagrangian can more easily incorporate the results of Einstein's special relativity. But while Dirac certainly had the key ideas in his mind, in his paper he merely developed a formalism that demonstrated useful correspondences and suggested a vague analogy between the action principle in classical mechanics and the more standard formulations of the evolution in time of a quantum mechanical wave function of a particle.

Feynman, being Feynman, decided right then and there to take some simple examples and see if the analogy could be made exact. At the time he was just doing what he thought a good physicist should do—namely, work out a detailed example to check what Dirac meant by what he said. But Jehle, who was watching this graduate student carry out his calculations in real time, faster than Jehle could follow, in that small room in the Princeton library, knew better. As he put it, "You Americans are always trying to find out how something can be used. That's a good way to discover things."

He realized that Feynman had carried Dirac's work one stage further, and in the process had indeed made an important discovery. He had established explicitly how quantum mechanics could be formulated in terms of a Lagrangian. In so doing, Feynman had taken the first step in completely reformulating quantum theory.

… … …

I ADMIT TO being skeptical about whether Feynman really outdid Dirac that morning in Princeton. Certainly anyone who understands Dirac's paper can see that almost all of the key ideas are there. Why Dirac didn't take the next step to see if they could actually be implemented is something we will never know. Perhaps he was satisfied enough that he had demonstrated a possible correspondence but never felt it would be particularly useful for any practical purposes.

The only information we have that Dirac never actually proved to himself that his analogy was exact is Feynman's later recollection of a conversation with Dirac at the 1946 bicentennial celebration at Princeton. Feynman describes asking Dirac if he knew that his "analogy" could be made exact by a simple constant of proportionality. Feynman's recollection of the conversation goes as follows:

Feynman: "Did you know that they were proportional?"
Dirac: "Are they?"
Feynman: "Yes."
Dirac: "Oh, that's interesting."

For Dirac, who was known to be both terse and literal in the extreme, this was a long conversation, and it probably speaks volumes. For example, Dirac married the sister of another famous physicist, Eugene Wigner. Whenever he introduced her to people, he introduced her as "Wigner's Sister," not as his wife, feeling apparently that the latter fact was superfluous (or perhaps merely demonstrating that he was as misogynistic as many of his colleagues at the time).

More relevant perhaps is a story I heard regarding the

famous Danish physicist Niels Bohr, who apparently was complaining about this far-too-quiet postdoctoral researcher, Dirac, who the equally famous physicist Ernest Rutherford had sent him from England. Rutherford then told Bohr a story about a person who goes into a pet shop to buy a parrot. He is shown a very colorful bird and told that it speaks ten different words, and its price is $500. Then he is shown a more colorful bird, with a vocabularly of one hundred words, with a price of $5,000. He then sees a scruffy beast in the corner and asks how much that bird is. He is told $100,000. "Why?" he asks. "That bird is not very beautiful at all. How many words then does it speak?" None, he is told. Flabbergasted, he says to the clerk, "This bird here is beautiful, and speaks ten words and is $500. That bird over there speaks a hundred words and is $5,000. How can that scruffy little bird over there, who doesn't speak a single word, be worth $100,000?" The clerk smiles and says, "That bird thinks."

WHAT DIRAC HAD intuited in 1933, and what Feynman picked up immediately and explicitly (although it took him awhile to describe it in these terms), is that whereas in classical mechanics the Lagrangian and the action function determine the correct classical path by assigning simple probabilities to the different classical paths between *a* and *c*—ultimately assigning a probability of essentially unity for the path of least action and essentially zero for every other path—in quantum mechanics the Lagrangian and the action function can be used to calculate, not probabilities, but probability *amplitudes* for transitions between *a* and *c*.

And that moreover in quantum mechanics many different paths can have nonzero probability amplitudes.

While working out this idea with a simple example Feynman discovered—before Jehle's surprised eyes that morning in the library at Princeton—that if he tried to calculate probability amplitudes using this prescription for very short travel times he could obtain a result that was identical with the result obtained in traditional quantum mechanics from Schrödinger's equation. What's more, in the limit where systems get big, so that classical laws of motion govern the system and quantum mechanical effects tend to become insignificant, the formalism that Feynman developed would reduce to the classical principle of least action.

How this happens is relatively straightforward. If we consider all possible paths between *a* and *c*, we can assign a probability amplitude "weight" to each path that is proportional to the total action for that path. In quantum mechanics many different paths—perhaps an infinite number, even crazy paths that start and stop and instantly change speeds and so on—can have nonzero probability amplitudes. Now the "weight factor" that is assigned to each path is expressed in terms of the total action associated with that particular path. The total action for any path in quantum mechanics must be some multiple of a very small unit of action called Planck's constant, the fundamental "quantum" of action in the quantum theory, which we earlier saw also gives a lower bound on uncertainties in measuring positions and momenta.

The quantum prescription of Feynman is then to add up all of the weights associated with the probability amplitudes for the separate paths, and the square of this quantity will determine the transition probability for moving from *a* to *c* after a time *t*.

The fact that the weights can be positive or negative accounts not only for the weird quantum behavior, but also for the reason why classical systems behave differently than quantum systems. For if the system is large, so that its total action for each path is then huge compared to Planck's constant, a small change in path can change the action, expressed in units of Planck's constant, by a huge amount. As a result, for different nearby paths the weight function can then vary wildly from positive to negative. In general, when the effects of these different paths are added together, the many different positive contributions will tend to cancel the very many negative contributions.

However, it turns out that the path of least action (the classically preferred path therefore) has the property that any small variation in the path produces almost no change in the action. Thus, paths near the path of least action will contribute the same weight to the sum, and will not cancel out. Hence, as the system becomes big, the contribution to the transition probability will be essentially completely dominated by paths very close to the classical trajectory, which will therefore effectively have a probability of order one, while all other paths will have a probability of order zero. The principle of least action will have been recovered.

WHILE LYING IN bed a few days later, Feynman imagined how he could extend the analysis he made for paths over very short time intervals to ones that were arbitrarily large, again by extending Dirac's thinking. As important as it was to be able to show that the classical limit was sensible, and that the mathematics could be reduced to the standard Schrödinger equation for simple quantum systems, what was most exciting for Feynman is that he now had a mechanism to explore the quantum mechanics of more complex physical systems, like the system of electrodynamics he devised with Wheeler, which they could not describe by traditional methods.

While his motivation was to extend quantum mechanics to allow it to describe systems that couldn't otherwise be described quantum mechanically, it is nevertheless true (as Feynman later emphasized) that for systems to which Dirac, Schrödinger, and Heisenberg's more standard formulations could be applied, all the methods are completely equivalent. What is important, though, is that this *new* way of picturing physical processes gives a completely different "psychological" understanding of the quantum universe.

The use of the word *picturing* here is significant, because Feynman's method allows a beautiful pictorial way of thinking about quantum mechanics. Developing this new way took awhile, even for Feynman, who did not explicitly talk about a "sum over paths" in his thesis. By the time he wrote up his thesis as an article six years later in *Reviews of Modern Physics*, the notion was central. That 1948 paper, titled "The Space-Time Approach to Non-Relativistic Quantum Mechanics," begins with a probability argument along the lines that I have given here, and then immediately starts discussing space-time paths. Surprisingly, drawings are conspicuously absent from the paper. Maybe it was too expensive in those days to get an artist to draw them. No matter, they would come.

... ...

WHILE FEYNMAN WAS writing up these results to form the basis of his thesis, the world in 1942 was in a state of turmoil, embroiled in the second world war of the century. Amid all of his other concerns—completing his thesis, getting married, finding a job—one morning he was suddenly interrupted in his office by Robert Wilson, then an instructor in experimental physics at Princeton. He sat Feynman down and revealed what should have been top-secret information, though the specific information was so new as to not yet have been thoroughly classified as such.

The United States was about to embark on a project to build an atomic bomb, and a group at Princeton was going to work on one of the possible methods for making the raw material for the bomb, a light isotope of uranium called uranium 235 (with the number 235 representing the atomic mass—the number of protons plus neutrons in the nucleus). Nuclear physics calculations had shown that the dominant naturally occurring isotope of uranium, uranium 238, could not produce a bomb with practical amounts of material. The question was, How can the rare isotope uranium 235, which could produce a bomb, be separated from the far more abundant uranium 238? Because isotopes of an element differ only in the number of neutrons in the nucleus, but otherwise are chemically identical since they contain the same number of protons and electrons, chemical separation techniques wouldn't work. Physics had to be employed. Wilson was revealing this secret because he wanted to recruit Feynman to help with the theoretical work needed to see if his own proposed experimental methods would work.

This presented Feynman with a terrible dilemma. He desperately wanted to finish his thesis. He was enjoying the problem he was working on, and he wanted to continue to do the science he loved. He also wanted to graduate, as this was one of his own preconditions for marriage. Moreover, Wilson would want him to focus on problems that Feynman viewed as engineering, a field he had explored but left for physics while still an undergraduate.

His first response was to turn down the offer. At the same time, how could he turn down a possibility to help win the war? He had earlier considered enlisting in the army if he could work in the Signal Corps, but he was told there were no guarantees. Here was a possibility of doing something far more significant. Moreover, he realized that the nuclear physics involved was not a secret. As he later said, "The knowledge of science is universal, an international thing. . . . There was no monopoly of knowledge or skill at that time . . . so there was no reason why if we thought it was possible that they [the Germans] wouldn't also think it was possible. They were just humans, with the same information. . . . The only way that I knew how to prevent that was to get there earlier so that we could prevent them from doing it, or defeat them." He did think for a moment about whether making such a frightening weapon was the right thing to do, but in the end he put his thesis work in a desk drawer and went to the meeting Wilson had told him about.

From that moment on he became occupied not in the abstract world of quantum mechanics and electrons, but in the minutiae of electronics and materials science. He was well prepared, as always, by what he learned on his own and in some excellent courses in nuclear physics from Wheeler and in the properties of materials from Wigner. Still, it took some getting used to. He and another research assistant of Wheeler's, Paul Olum, a Harvard graduate in mathematics, set to work as fast as they could doing calculations they were not certain about even as the experimentalists around them were building the device that the two of them had to determine was or was not workable.

This was Feynman's first experience with the realization, which would reoccur many times over his subsequent career, that while he loved theoretical calculations, he didn't really trust them until they were put to experimental tests. As daunting as it was to try to understand nature at the edges of knowledge, it was equally daunting to be responsible for decisions based on his calculations that would ultimately have an immediate impact on the largest industrial project ever carried out by one country.

In the end, Wilson's proposed device for isotope separation was not chosen. What was selected for major production of U235 were the processes known as electromagnetic separation and gaseous diffusion.

Feynman's thesis supervisor, John Wheeler, did not abandon him during this period. Wheeler had left Princeton to work with Enrico Fermi on building the first nuclear reactor at the University of Chicago—where they would test the principle of controlled chain reactions as a first step toward the uncontrolled chain reactions that would be required to build a nuclear bomb. But Wheeler was aware of what Feynman was working on, and in the spring of 1942 he decided enough was enough. He and Wigner felt that Feynman's thesis work was close enough to completion to be written up, and he told him so, in no uncertain terms.

Feynman proceeded to do just that. He was aware of what he had achieved. He had re-derived quantum mechanics in terms of an action principle involving a sum (or rather, in the language of mathematics, an integral) over different paths. This allowed a generalization to situations where the standard Schrödinger approach would not work—in particular the absorber theory that he and Wheeler had worked out for electromagnetism. This is what interested him—his real step forward, he thought—and the new method he had invented for deriving quantum mechanics was primarily a means toward this end.

But he was more concerned about what he hadn't yet achieved, and he devoted the final section of his thesis to describing the limitations of his work thus far. First and foremost, his thesis did not contain any comparison with experiments, which he regarded as the real test of the worth of any theoretical idea. Part of the problem was that while he had reformulated purely nonrelativistic quantum mechanics, he was acutely aware that in order to address real experiments involving charges and radiation, the appropriate theory—quantum electrodynamics—was needed so as to incorporate relativity, which involved addressing a host of problems he had not yet dealt with.

Finally, Feynman was concerned with the physical interpretation of his new viewpoint for dealing with the quantum world—in particular, the issue of making a connection between the temporally spread-out paths and probability amplitudes inherent in his new formulation, and the possibility of making real physical measurements at any specific time. The problem of measurement was not new or unique to Feynman's thesis. His work merely appeared to exacerbate it. The world of measurements lies within the classical world of our experience, where weird quantum paradoxes don't ever seem to arise. How does a "measurement" ensure that the underlying quantum universe ends up appearing sensible to our eyes?

The first person to comprehensively attempt to quantitatively discuss this measurement problem in the context of quantum mechanics was John von Neumann, at Princeton, whom Feynman had the opportunity to interact and disagree with. Anyone who has heard anything about quantum mechanics often hears that one cannot separate the observer from that which is being observed. But in practice this is exactly what is required in order to make predictions and compare them with experimental data. Feynman was particularly concerned about this key question of how to separate the measuring apparatus and the system being observed in the context of the specific quantum mechanical calculations he wanted to perform.

The conventional verbiage goes as follows: When we make a measurement, we "collapse the wave function." In other words, we suddenly reduce the probability amplitude to be zero in every state but one. Therefore, the system has a 100 percent probability of being in only one configuration, and different possible configurations do not interfere with each other, as in the examples discussed in the last chapter. But this simply begs the questions: How does a measurement collapse the wave function, and what is so special about such a measurement? Is a human being needed to make the observation?

New-age hucksters aside, consciousness is not the key. Rather, Feynman argued that we must consider the system plus the observer together as a single quantum system (which is fundamentally true, after all). If the observing apparatus is "large"—that is, it has many internal degrees of freedom—then we can show that such a large system behaves classically—interference between different possible macroscopic quantum states of the apparatus becomes infinitesimally small, so small as to be irrelevant for all practical purposes.

By the act of measurement, we somehow produce an interaction between this "large" observing system and a "small" quantum system, and these become correlated. This correlation ultimately fixes the small quantum mechanical system to also exist in a single well-defined state, the state we then "measure" the system to be in. In this sense, we say the wave function of thc small system has *collapsed* (meaning the probability amplitude of being in any state other than the one we measure is now zero). Humans have nothing to do with it. The observing system simply has to be large and classical and correlated with the quantum system via a measurement.

This still does not fully resolve the question, which then becomes: how do we determine what comprises the large classical observing part of the combined system and what comprises the quantum part? Feynman spent considerable time discussing this issue with von Neumann. He was not satisfied with von Neumann's argument that someone had to decide, in some sense arbitrarily, where to make the cut between classical observer and observed. That sounded like a philosophical cop-out. Feynman believed that since quantum mechanics underlies reality, it should be consistently incorporated throughout instead of making ad hoc separations between an observer and the observed. In fact, he worked hard to define measurements purely in terms of correlations between different subsystems and letting the size of one of them go to infinity. If nonzero and finite correlations remained in this limit, Feynman labeled this a "measurement" of the smaller subsystem, which could be made arbitrarily accurate as the size of the "measuring" part of the system grew to be larger and larger. As he colorfully put it in a note he wrote to himself, regarding the example of a spot on a photographic plate that somehow recorded an event involving a single atom,

> What can we expect to end with if we say we can't see many things about one atom precisely, what in fact can we see. Proposal: Only those properties of a single atom can be measured which can be correlated (with finite probability) (by various experimental arrangements) with an unlimited number of atoms. (i.e. the photographic spot is "real" because it can be enlarged and projected on screens, or affect large vats of chemicals, or big brains etc etc—it can be made to affect ever increasing sizes of things—it can determine whether a train goes from N.Y. to Chicago—or an atom bomb explodes—etc).

Measurement theory still remains the bugaboo of quantum mechanics. While great progress has been made, it is still fair to say that a complete description of how the classical world of our experience results from an underlying quantum reality has not been developed, at least to the satisfaction of all physicists.

This example of Feynman's focus when finishing his thesis is important because it demonstrates the sophisticated issues this mere graduate student in physics insisted on wrestling with as he worked. In addition, Feynman's "path-integral formalism" made it possible to separate systems into pieces, which seems central to the idea of measurement in quantum mechanics, allowing one to isolate parts of a system one either does not or cannot measure and to separate these clearly from those parts one wishes to focus on. This is generally not possible in standard formulations of quantum mechanics.

The idea is really relatively straightforward. We sum all those weights corresponding to the action associated with those paths or parts of paths we wish to ignore the specific details of, for example, summing up the effect of having small circular loops—so small we could never measure them—swirling around the more normal straight trajectories between two points. The effect of these additions may be to change, by a small calculable amount, what would be the action associated with a straight trajectory without the loops. After having done the summation (or in the case of an infinite number of such additional paths, the integral), we can then forget about such extra loop trajectories and focus on only trajectories that are more straight, as long as we use the new, altered action for this trajectory in our calculations. This process is called *integrating out* parts of the system.

This may seem, at first sight, like a technical detail not worth mentioning. However, as we will see later, it ultimately would allow almost all of the most important theoretical advances in fundamental physics in the twentieth century to occur, and it would allow us to totally quantitatively revolutionize what are otherwise such vague notions as scientific truth.

But for the moment, in 1942 as he completed this thesis, titled "The Principle of Least Action in Quantum Mechanics," Feynman had other things on his mind. Preparing to graduate that June, he had received word that he was to move to Los Alamos to focus on the actual building of the atomic bomb. He was also busy planning for his long hoped-for marriage following his graduation. He thus had to put his immediate physics concerns to rest and focus on sorting out the rest of his life. Perhaps all of the diversions explain why, even as he acknowledged Professor Wheeler for his advice and encouragement, he never took the opportunity to add what surely would have been a far more poetic acknowledgment to the connection made between the subject of the thesis (and ultimately the work that would win him the Nobel Prize) and that fateful afternoon in his high school physics class when Mr. Bader awakened his mind to the subtle beauties of theoretical physics.

Three years and what undoubtedly seemed like a lifetime later, the war had ended and he finally got around to writing up his thesis for publication. He still did not make this connection. Instead he was able to clearly enunciate what undoubtedly had been the very same hopes he carried with him as he left Princeton, and which had continued to buoy him through the various immediate insanities of the world of human affairs over which he had so little control, until the day he was finally free to explore full-time the more intoxicating insanities of the quantum universe, which he felt much more confident he could conquer:

> The formulation is mathematically equivalent to the more usual formulations. There are therefore no fundamentally new results. However there is pleasure in recognizing old things from a new viewpoint. Also there are problems for which the new viewpoint offers a distinct advantage. . . . In addition, there is always the hope that the new point of view will inspire an idea for the modification of present theories, a modification necessary to encompass present experiments.

CHAPTER 6

Loss of Innocence

He's another Dirac. Only this time human.

—EUGENE WIGNER, SPEAKING ABOUT RICHARD FEYNMAN

Richard Feynman graduated with a PhD from Princeton in 1942 as a relatively naive and hopeful young man, known to his fellow students and professors as a brilliant and brash intellect, but largely unknown outside the university. He emerged three years later, from Los Alamos, as a well-tested physicist highly regarded by most of the major players in physics around the world, and a somewhat jaded and world-weary adult. Along the way, he experienced incredible personal loss, as well as the loss of intellectual and moral innocence that is the inevitable by-product of war.

THE INK HAD barely dried on Feynman's diploma when he began to execute his decision, outlined in that dispassionate letter to his mother, to marry Arline. The opposition by his parents and Arline's, who were more concerned about his health and Arline's than about their mutual love, was futile. Both he and Arline felt the other was a bastion against any onslaught from the rest of the world. Together anything was possible, and they refused to be pessimistic about the future. As Arline wrote to Richard shortly after he had moved into a new flat in Princeton and made final arrangements for the ceremony, "We're not little people—we're giants. . . . I know we both have a future ahead of us—with a world of happiness—now and forever."

Every aspect of their brief lives together is, in retrospect, heart wrenching. On what would be their wedding day, Richard borrowed a station wagon from a friend, which he outfitted with mattresses so Arline could lie down. Then he drove from Princeton to her parents' home and picked her up in her wedding dress, and together they drove to Staten Island for a wedding ceremony with no family or acquaintances, and from there to what would become Arline's temporary new home, a charity hospital in New Jersey.

Shortly thereafter, without any fanfare or honeymoon, Feynman returned to work at Princeton, except there was nothing to do. The project with Wilson had been closed down and the team was waiting for new orders. Since the main activity at the time was taking place in Chicago, where Enrico Fermi and Wheeler were working on building a nuclear reactor, Feynman was sent to Chicago to learn what was going on there.

His trip in 1943 began what would be a succession of opportunities to ultimately meet and impress his peers and his bosses. While the war disrupted all lives, in at least two senses it provided Feynman with incredible opportunities he would not otherwise have had.

First, since the best and brightest minds were being

gathered to spend two years in close quarters, Feynman was given a chance to shine in front of individuals he would have otherwise had to travel around the world to meet. He had already, through his attendance since 1942 at periodic group meetings in New York and at the MIT Radiation Laboratory in Massachusetts, impressed the brilliant, if troubled, physicist Robert Oppenheimer, who would shortly be chosen to lead the entire atomic bomb project. In Chicago, while carrying out his job of gathering information, he blew away members of the theory group there when he was able to perform a calculation that had eluded them for over a month.

Following his return to and debriefing at Princeton, he didn't have to wait long before learning what was to come next. Oppenheimer had been chosen to lead the bomb project, and shortly after that he picked Los Alamos, New Mexico—a remote and starkly beautiful countryside where he had previously roamed as a younger man, and which also fit the army's requirements of isolation and safety—as the site of what would soon become the most advanced laboratory in the world, with the highest concentration of brilliant scientists ever seen per square mile (even allowing, as John F. Kennedy once did, for those days when Thomas Jefferson dined alone in the White House).

Oppenheimer was a brilliant scientist, but more important for the success of the atomic bomb project, he was an equally brilliant judge of talent in others. He quickly began to recruit and amass a team of outstanding colleagues to relocate to Los Alamos even before the laboratory and associated housing had been completed. Needless to say, he sought out Feynman, and did whatever he could to convince him to move to New Mexico with the first wave of scientists, at the end of March in 1943.

Oppenheimer's offer led to the other fortuitious impact that the war effort had on the married couple. Arline's illness was progressing. She would live only two years following their marriage. The first years of any marriage should be a time, if there is ever going to be a time, of romance and adventure. Had the war not turned the world topsy-turvy, Feynman undoubtedly would have taken longer to finish his doctorate, he and Arline would have continued their strained existence in Princeton as her health deteriorated, and then, before she died, he might have proceeded to an assistant professorship in some place not that different than Princeton. Instead, his decision to move to the wild and unknown Southwest would give the young couple, especially Arline, the chance for a morsel of the romance and adventure that they had been longing for and that she otherwise would never have been able to enjoy.

Feynman was touched by Oppenheimer's concern and consideration. "Oppie," as he was known to his colleagues, seemed to be the perfect leader for this group of independent-minded scientists. He commanded their respect—as Feynman later said, "We could discuss everything technically because he understood it all." At the same time he showed uncommon concern about the well-being of each and every person he had recruited for this task. Again, as Feynman remembered it, "Oppenheimer was extremely human. When he was recruiting all these people to go to Los Alamos . . . he still worried about all the details. For example, when he asked me to come I told him I had this problem—that my wife had tuberculosis. He himself found a hospital and called me up to say they had found somewhere that would take care of her. I was only one of all the many people he was recruiting, but this was the way he always was, concerned with people's individual problems." Oppenheimer's call from Chicago about finding a hospital for Arline was the first telephone call Feynman had received from so far away, perhaps one of the reasons he was so touched. In any case, after some negotiations with the army authorities, Arline and Richard were set to board the Santa Fe "Chief" from Chicago on March 30. Arline was beside herself with joy and excitement:

> Dearest Rich—if you only knew how happy you've made me with this train trip of ours—it's all I've wanted and dreamed about since we've been married . . . with only one day left—I'm so excited and happy and bursting with joy—I think, eat, and sleep "you"—our life, our love, our marriage—the great future we are building. . . . If only tomorrow would hurry and come.

At Arline's urging, the two of them purchased a ticket for a private room, then they boarded the train and headed west. Ultimately, after exploring several possibilities, Arline was placed in a sanatorium in Albuquerque, one hundred miles from the laboratory site (there was no laboratory there yet), and Richard somehow made the trip to see her once a week.

In one sense, Richard Feynman had been preparing for this experience his whole life. All of his talents were to be exploited during the next two years: his lightning computational abilities, his mathematical wizardry, his physical intuition, his clear appreciation for experiment, his disrespect for authority, his breadth of physics knowledge from nuclear physics to the physics of materials (shortly after arriving he became ill, and in a letter to his mother reported reading a chemical engineering textbook with topics ranging from "Transportation of Fluids" to "Distillation" while in the infirmary for three days), and his fascination with computing machines.

The physics work was quite different from his academic work. It was easier than pushing the forefront into the unknown laws, but a lot dirtier than working on pristine single electrons in hydrogen atoms. Aside from his contributions to the development of the bomb, Feynman left little permanent scientific legacy from his work during this time. (There is a formula for the efficiency of a nuclear weapon, called the Bethe-Feynman formula, that is still used today, but that is about it.)

Nevertheless, Los Alamos had a profound influence on Feynman's career, and it all began by accident, as so many things do. Again, in his words: "Most of the big shots were out of town for one reason or the other, getting their furniture transferred or something. Except for Hans Bethe. It seems that when he was working on an idea he always liked to discuss it with someone. He couldn't find anybody around, so he came down to my office . . . and he started to explain what he was thinking. When it comes to physics I forget exactly who I'm talking to, so I was saying, 'No, no! That's crazy!' and so on. Whenever I objected, I was always wrong, but nevertheless that's what he wanted." As Bethe remembers it, "I knew nothing about him. . . . He had only recently got his Ph.D. from Wheeler, at Princeton. We got to talking, and he obviously was very bright. At the meetings and seminars he always asked questions which seemed particularly intelligent and penetrating. We began to collaborate together." And in another reminiscence, "He was very lively from the beginning. . . . I realized very quickly that he was something phenomenal. . . . I thought Feynman perhaps the most ingenious man in the whole division, so we worked a great deal together."

The opportunity to work with Bethe at Los Alamos was fateful in the extreme. They complemented each other in remarkable ways, sharing uncanny physical intuition, mental stamina, and calculational ability. But Bethe was, in several other senses, everything that Feynman was not. He was calm and deliberate, and unlike the excitable Feynman, Bethe was "unflappable." This was also reflected in their mathematical styles. Bethe began a calculation at the beginning, and ended at the end, no matter how long or difficult the road between was. Feynman, on the other hand, was as likely to begin in the middle or at the end, and jump back and forth until he had convinced himself he was right (or wrong). In other areas, Bethe served as a remarkable role model. Feynman loved his humor, his unaffected manner, and his straightforward and collegial way of dealing with others. And whereas Wheeler helped free up Feynman's enthusiasm and creativity, he was not the physicist that Bethe was. If Feynman was to rise to new, and higher levels, he needed someone he could go head to head with. Bethe was the man.

By the time Bethe had moved to Los Alamos, he had resolved one of the most important and vexing questions in astrophysics: how does the sun shine? For over a century scientists had wondered what energy process powers the sun so it has been able to shine with its observed luminosity for over 4 billion years. The earliest estimate, by a German doctor in the early eighteenth century, suggested that if the sun were a big ball of burning coal, it could burn with its observed brightness for about 10,000 years, which happened to be in nice accord with some biblical estimates of the age of the universe. Later in the century, two famous physicists, Heinrich Helmholtz and Lord Kelvin, estimated that the sun could be powered by the energy released during gravitational contraction, and this energy source could power the sun for perhaps 100 million years. However, even this estimate was far too low to explain what was by then the inferred age of the solar system—namely, billions, not hundreds of millions, of years.

The mystery persisted through the 1920s, when the famous British astrophysicist Sir Arthur Stanley Eddington argued that there must be some unknown source of energy powering the solar interior. The problem was that model calculations of the sun's profile suggested that the interior was no more than 10 million degrees in temperature, which is hot, but not that hot. In other words, the physical processes associated with the energies available at these temperatures were thought to be fairly well understood, with no room for new exotic physics. As a result, Eddington's assertion was met with skepticism, leading him to utter his famous rebuke: "To those that think the temperature in the center of the Sun is not hot enough for some new physical process to take place, I say: Go and find a hotter place!"

Bethe, who had studied with the greatest theoretical physicists in Europe, including Arnold Sommerfeld, Paul

Dirac, and Enrico Fermi, had, by the early 1930s, established himself as perhaps the world's foremost authority on the emerging field of nuclear physics. He wrote the definitive set of reviews in this field, which Feynman had studied while an undergraduate. If anyone was prepared to find the new process that powered the sun, it was Bethe, and in 1939 he made his great discovery. He realized that newly discovered nuclear reactions (similar in spirit to those later exploited in building the fission bomb, but instead of being based on breaking up heavy nuclei such as uranium and plutonium, these involved fusing light nuclei such as hydrogen into heavier nuclei) provided the key to releasing tremendous amounts of energy. Moreover, he showed that there was a series of reactions starting with protons, which make up the nuclei of hydrogen, and ultimately producing the nuclei of the next lightest element, helium, that would release more than twenty million times as much energy as comparable chemical reactions between hydrogen would release. While at a temperature of only 10 million degrees the average hydrogen nucleus might take over a billion years to experience a collision energetic enough to initiate such a reaction, over a hundred thousand tons of hydrogen could nevertheless convert to helium each second, providing enough energy to power the sun at its current luminosity for about 10 billion years.

For this important theoretical discovery, Bethe was awarded the Nobel Prize in Physics in 1969, four years after Feynman would share the prize for his own work on quantum electrodynamics (QED). And the nuclear "fusion" reactions Bethe exploited in his explanation of the workings of the sun would be re-created seven years after the end of World War II, in the development of "thermonuclear explosives," otherwise known as hydrogen bombs.

Oppenheimer had recruited Bethe in 1942 and wisely chose him to head the Theoretical Division, which contained the biggest brains and the biggest egos that would reside at Los Alamos. Not only was Bethe their intellectual equal, but also his calm yet persistent strength of character would be essential in helping to guide them, put out fires, and, above all, put up with their idiosyncrasies.

In Feynman, Bethe had found just the foil he needed to bounce ideas off of, just as Feynman had found the perfect mentor to help steer his active imagination. That they both loved their work didn't hurt either. Bethe, to his credit, recognized Feynman's talent quickly and made what might seem the unorthodox decision to name the twenty-four-year-old a group leader in the Theoretical Division, outranking colleagues older and more experienced. Stephane Groueff recalled their interactions: "Richard Feynman's voice could be heard from the far end of the corridor: 'No, no, you're crazy!' His colleagues in the Los Alamos Theoretical Division looked up from their computers and exchanged knowing smiles. 'There they go again!' one said. 'The Battleship and the Mosquito Boat!' "

It is not hard to guess who was who. Nevertheless, beyond the hearty joint laughter and intellectual jousting, what left the most lasting impression on this still impressionable young man was Bethe's insisting on connecting every theoretical calculation with a number, a quantity that could be compared with experimental results. It is hard to overemphasize how deeply this governed almost everything Feynman did in later life as a scientist. As he later put it,

"Bethe had a characteristic which I learned, which is to calculate numbers. If you have a problem, the real test of everything—you can't leave [it] alone—you've got to get the numbers out; if you don't get down to earth with it, it really isn't much. So his perpetual attitude is to use the theory. To see how it really works is to really use it."

The catalogue of the activities Feynman accomplished while working under Bethe at Los Alamos was remarkable, not least for their diversity. He began by quickly developing a method to numerically integrate (or sum) so-called third-order differential equations, which had derivatives of derivatives of derivatives in them. His method turned out to be more accurate than what one could do with simpler second-order equations. Less than a month later, Feynman and Bethe had worked out their formula for calculating the efficiency of a nuclear weapon.

He then moved on to the more theoretically challenging problem of calculating the diffusion of fast neutrons that triggered fissions in the uranium 235 atomic bomb. For this problem he developed an approach that was very similar mathematically to the formulation he would eventually create for dealing with QED.

During the final phases of building the bomb, Feynman was put in charge of computing, ultimately supervising all computational aspects of assembling a successful plutonium bomb, which John von Neumann had suggested could be triggered by a massive implosion, increasing the density of material and making an otherwise stable mass go critical. The first human-induced nuclear explosion, above the desert floor just before sunrise on July 16, 1945, code-named Trinity, was successful in no small part because of

Feynman's calculational leadership in these crucial last months.

Feynman's work involved using and even assembling a new generation of electromechanical computing machines to perform the complex modeling calculations necessary to design the new device, which challenged Feynman's mechanical as well as his mathematical prowess. As Bethe later described it,

> Feynman could do anything, anything at all. At one time, the most important group in our division was concerned with calculating machines. . . . The two men I had put in charge of these computers just played with them, and they never gave us the answers we wanted. . . . I asked Feynman to take over. As soon as he got in there, we got answers every week—lots of them, and very accurate. He always knew what was needed, and he always knew what had to be done to get it. . . . (I should mention that the computer had arrived in boxes—about ten boxes for each. Feynman and one of the former group leaders put the machines together. . . . Later we got some professionals from IBM who said, "This has never been done before. I have never seen laymen put together one of these machines, and it's perfect!"

The degree to which Feynman contributed to the successful development of the bomb as he exploited his natural talents and matured as a physicist was well described by the physicist and historian of science Sylvan Schweber: "His versatility was legendary. His genius at lockpicking, repairing Marchant and Monroe calculators, assembling IBM machines, solving puzzles and difficult physics problems, suggesting novel calculational approaches, and explaining theory to experimenters and experiments to theoreticians earned him the admiration of all those with whom he came in contact."

Feynman's talents and energies at Los Alamos derive from a characteristic his old college chum Ted Welton, who later came to Los Alamos to work with him, described: "Once presented with a clearly formulated physical paradox, mathematical result, card trick, or whatever [Feynman] would not sleep until he had the solution." Schweber agreed, saying this comment captured the quintessential Feynman, who had "an obsessive need to 'undo' what is 'secret.'"

Feynman's accomplishments were all the more impressive considering that in the midst of all this, his wife lay dying in the Albuquerque hospital. Every week, he made the 200-mile round-trip to visit her, either borrowing cars or hitchhiking all the way. His correspondence with her increased as her symptoms worsened, becoming almost daily near the end. Their mutual love, and the tenderness and concern he expressed for her are obvious, and painful to read.

In the four months before Arline died, on June 16, 1945—six weeks before the first atomic bomb was dropped over Hiroshima—Richard Feynman wrote Arline thirty-two letters. He wrote to doctors to explore and solicit new treatments for tuberculosis, and moved her to the Los Alamos site so they could be closer together, until her unhappiness with the army nurses, the regulations, and the living arrangements caused Richard to move her back to Albuquerque, in spite of his own misgivings. He wrote to her about his regrets about getting drunk on VE Day, about their joint fear that she might be pregnant, about packages from home, about fighting a forest fire, and about men being banned from the girls' dorm (he joked that he hadn't been in a girls' dorm for over a year), but most of all he wrote about his love for her. The very last letter he wrote to her, on June 6, ended as follows:

> I will come this week and if you don't want to bother to see me just tell the nurse. I will understand darling, I will. I will understand everything because I know now that you are too sick to explain anything. I need no explanations. I love you, I adore you, I shall serve you without question, but with understanding. . . . I adore a great and patient woman. Forgive me for my slowness to understand. I am your husband. I love you.

Meanwhile, during this impossible time, he and the other scientists at Los Alamos were proceeding at a frenzied pace toward the creation of a bomb that would change history forever. Emotions were high, and maybe that is what kept them going. When Germany was defeated, no one seemed to ask the question, Why are we building this bomb? Everyone wanted to see the fruit of their intense labors brought to light, to end the war in the Pacific.

To someone like me, who grew up in the era of big science and big bureaucracy, the remarkable intensity and speed with which the Manhattan Project was carried out are almost unfathomable. The time from when the bomb was mere theoretical speculation to the Trinity test was less than five years. The time from when Feynman and others were recruited was less than three years. The best physics minds, initially motivated by a belief that the enemy (Germany) was also developing a nuclear weapon, accomplished in three years what in the current world might take at least a decade or two. The mammoth isotope separation project at the Oak Ridge laboratory in Tennessee alone, which magnified by more than a millionfold the total amount of separated uranium 235 in the world during that three-year period—under very dangerous circumstances that Feynman was sent, in 1945, to try to correct—would probably take at least as long in the current world just to get environmental clearance before being allowed to begin.

Feynman had rushed to be by Arline's side during her last hours on June 16, but after she died, he realized that the dead no longer need help, so he gathered her possessions, arranged for her immediate cremation, and amazingly returned to Los Alamos, shattered as he was, to report to work the next evening. Bethe would have none of this and ordered him to go home to Long Island for a rest. His family had no advanced warning that he was arriving, and he stayed until he received a coded telegram almost a month later summoning him back to New Mexico. He returned on July 15, was driven to Bethe's house for sandwiches made by Bethe's wife, Rose, and then caught a bus to a desolate desert location called Jornada del Muerto, where he would join his colleagues to watch the test of the device they had spent day and night for the last three years designing and building, the device that would change the world forever.

Everyone who saw the blast was awed, but differently. Some, like Oppenheimer, recalled poetry, in his case an obscure passage from the Bhagavad Gita: "Now I am become Death, the destroyer of worlds." Feynman, who had managed to avoid superstition at the moment of his wife's death, and sentimentality immediately afterward, held true to form. He thought about the processes that formed the clouds around the blast wave, and the processes that caused the air to glow as it became ionized in the heat of the explosion, and a hundred seconds later, by the time the sonic boom from the blast finally arrived at the observation deck, he was grinning. The calculations he had worked so hard on had been validated by nature.

CHAPTER 7

Paths to Greatness

There is pleasure in recognizing old
things from a new viewpoint.

—RICHARD FEYNMAN

It was the best of times. It was the worst of times. Richard Feynman left Los Alamos in October of 1945 as a shining new star in the physics community. As early as 1943 Oppenheimer had tried to convince the chairman of the Physics Department at the University of California, Berkeley to offer him a position, saying, "He is by all odds the most brilliant young physicist here . . . his excellence is so well known, both at Princeton . . . and to a not inconsiderable number of 'big shots' on this project, that he has already been offered a position for the post war period, and will most certainly be offered others." Oppenheimer was referring to an offer from the ever-astute Bethe, who by November of 1943 had arranged a faculty offer for Feynman at Cornell University such that Feynman was officially on leave from Cornell while at Los Alamos. The staid Berkeley chairman delayed making an offer until the summer of 1945, telling Feynman, "No one to whom we made an offer ever refused it." But Feynman did. He had come to know and love Bethe, who had been building an excellent group at Cornell. Moreover, Bethe managed to get Cornell to counter Berkeley's offer and beat its salary, so Feynman left Los Alamos in the fall, bound for Cornell, the first of the group leaders to leave the site.

Oppenheimer's prediction was right on the money, as it so often was. Within a year, Feynman had received further offers of permanent positions at the Institute for Advanced Study, Princeton University, and UCLA, all of which he turned down so he could stay with Bethe's group. The offers led to his being promoted at Cornell from assistant to associate professor.

All of these endorsements should have buoyed Feynman's spirits. However, he was worn out, pessimistic, and depressed. Even though it had been expected, the death of his wife must have been a terrible blow. Arline had been his lifeline. Compounding the depression he must have felt following her death was a more general sense of anomie brought on by the knowledge that nuclear weapons could be built, and that the United States would not have a monopoly on them for long. Feynman later remembered, for example, being with his mother in New York immediately after the war and thinking about how many people would be killed if a bomb were dropped there.

At that point, contemplating the future seemed pointless because he felt there was no future. Like Einstein, who said, "Everything has changed, save our way of thinking," Feynman saw no change in international relations after the war, and he was certain nuclear weapons would be used again soon. As he said, "What one fool can do, another can."

He thought it was silly to build bridges that would soon be destroyed. So too, he must have thought, Why try to bridge a new understanding of nature?

In addition to all of this was the natural letdown after the intense pace and pressure of participating in the grand and intense Manhattan Project: the intellectual challenge, the rapid gratification, the teamwork . . . all of that changed when he moved to Cornell. After the remarkably productive war years, in which real problems had to be urgently solved, and the results put to practical tests in short order, the return to pondering questions where progress is inevitably slower and more diffuse must have been even more difficult for Feynman.

The problems associated with the atomic bomb project may have been mathematically challenging, but they essentially involved well-understood physics or engineering. In this sense they were like solving problems in class—both well defined and easy—except the stakes were much higher and the pressure more intense. The problems Feynman was returning to involved deep questions of principle, where no one knew the right direction. He could work for years with no apparent progress. This kind of research can be discouraging under the best of conditions.

Combined with this was his concern that he had wasted three years of his life when he could have been attacking these problems, and a worry that the world was passing him by. Julian Schwinger, whom he first met at Los Alamos shortly after the Trinity test, and with whom he would later share the Nobel Prize, was the same age as Feynman, twenty-seven, but Schwinger already had his name attached to discoveries in physics (and within two years would be named a full professor at Harvard), whereas Feynman felt he had nothing to show for all of his efforts.

Last, there was the shock of suddenly beginning a new career as a university professor. After a focus on research problems both deep and otherwise, all of the minutiae that new instructors are bombarded with, and for which they have almost no preparation, can be disabling to say the least. Feynman arrived back at Cornell earlier than his other colleagues. Bethe didn't arrive until December, and so he was not available to mentor Feynman through the transition to his new job.

Teaching takes far more time than one expects, and being a successful teacher, which Feynman apparently was in mathematical physics and electricity and magnetism, can create a negative feeling about the state of one's research. Einstein once said that teaching is good because it gives the illusion each day of having accomplished something. A good lecture can provide immediate fulfillment, whereas research can go on for months at a time without making progress.

On top of this, Richard's father—the man who had first encouraged him to solve puzzles and prodded him to enjoy questioning nature, who had been so concerned about his son's future, and who had written him a letter finally swelling with pride after his appointment at Cornell—died suddenly of a stroke one year after Arline died and at the height of Feynman's depression. His father's pride only exacerbated his concerns about his own accomplishments. He realized that his last published paper was the one he wrote while he was an undergraduate at MIT, leaving him with the sense that he was burned out, that, at the tender age of twenty-eight, his best years were behind him. This attitude made the rush of attractive and ever-more lucrative job offers seem even worse, as if he didn't deserve the adulation and he would no longer accomplish anything of importance.

It can be very demoralizing to be viewed by the world differently than one feels about oneself. I remember again, from my own personal experience how, after I received my dream fellowship from Harvard, I couldn't focus on work for at least three months because after five years in the same city as an anonymous graduate student, the transition in status made me feel unworthy at best. When Feynman received the simultaneous offers from Princeton and the Institute for Advanced Study, his response was, "They were absolutely crazy."

Feynman credited a number of things with getting him out of his funk, including a talk with Robert Wilson, who had moved to Cornell to direct its new nuclear studies laboratory. Wilson told him to stop worrying, that he should not feel any pressure; his hire was Cornell's risk, not his. Feynman later related a famous story that reminded him why he enjoyed doing physics: he saw a plate thrown up in the cafeteria wobble strangely once every two revolutions, and decided to figure out why it moved, just for fun.

Probably more contributory to his recovery was the passage of time. He needed to get over his wife's and father's deaths and to reconcile with his mother—whose antagonism toward his decision to marry Arline had somewhat estranged her from her son—and to regain the rhythm of physics as he had known it before the war. His character, which normally radiated enthusiasm and self-confidence, and his mind, which had been focused toward the grand adventure of solving nature's puzzles, could not be submerged forever. (Bethe later remarked, on learning about Feynman's depression, which was not obvious to others at Cornell, "Feynman depressed is just a little more cheerful than any other person when he is exuberant.")

DURING THE WAR years, Richard had not completely stopped thinking about his physics problems. He carried around scraps of paper with calculations, often performed during the weekly trips to see Arline, where he would return to the question of how to formulate a real quantum theory of electromagnetism. Primarily he focused on the question of how to properly incorporate Einstein's special relativity into his equations.

Recall that he had learned, even as an undergraduate, that none other than Paul Dirac had discovered the proper way to describe the relativistic motion of electrons, and it was Dirac's paper on the Lagrangian formulation of quantum mechanics that had inspired the work leading to Feynman's thesis. The problem was that while Feynman's "sum-over-paths" approach could reproduce completely Schrödinger's equation, appropriate for nonrelativistic quantum mechanics, Feynman could not find a way to easily extend it to reproduce the Dirac equation in a relativistic context. He found that when he calculated energies he kept getting answers that were nonsensical, involving the square roots of negative numbers, and furthermore when he tried to calculate probabilities, summing up the probabilities of all events would not yield 100 percent.

After the war, as he began to return to these issues, he

focused first on what should have been the easier task: writing the results of his 1942 PhD thesis, which still had not been formally published. Here another character trait of Feynman began to clearly emerge. He did not easily write up his work for publication. He had no problems writing down his results in a colloquial form for his own use and for his own understanding—in fact, he did this on numerous occasions with great discipline—but to write for publication required a formal didactic format in which he needed to give not so much a history of how he had figured something out but rather a logically coherent step-by-step analysis of the final results using language and relations that the rest of the community would be comfortable with.

Moreover, there was the other issue of getting things exactly right. When Feynman was working through problems on his own, answers were often intuited, and then he could check afterward if he was right by trying many specific examples. He didn't follow any clear chain of logic, and yet he was aware that this was what was required for published articles. Translating his results took great effort—for Feynman, it was worse than pulling teeth.

His work ultimately appeared in 1947 in the journal *Reviews of Modern Physics* after it was rejected by the more conventional research journal *Physical Review*. However, it might not have appeared in that publication that year if his friends Bert and Mulaika Corben, with whom he visited in the summer of 1947, hadn't forced him to write the paper during his stay. Bert described it simply as follows: "We practically locked Dick in a room and told him to start writing." The story grew over the years, and by the time the physicist Freeman Dyson—who himself said of Feynman that "it took extreme measures to persuade him to write anything!"—related the story in his memoirs, it was centered on Mulaika and was more extreme: "She got Feynman into her house, and simply locked him up in a room and refused to let him out until he'd written the paper. I think she even refused to feed him unless he wrote it."

Whatever actually happened, reworking his thesis allowed him not only to refresh his old ideas but also to extend them so that his formulation of quantum mechanics began to be more visual. He started to "think" in terms of paths. Indeed, in his article, Feynman for the first time explicitly describes quantum mechanics in this new language—in terms of a "sum over paths." As he later said, "Clarity came from writing up the *RMP* article. . . . I could see the path . . . each path got an amplitude." With this work, Feynman completed his reformulation of our understanding of quantum mechanics. The true significance of the reformulation, and the recognition that at some deep level it might be not only more fundamental than conventional pictures, but also much more powerful, would take time to sink in, for both Feynman and the rest of the physics community.

FREED FROM THIS burden, Feynman then returned to the task of trying to formulate a relativistic quantum theory of electromagnetism. He tackled this one as he did all such problems, by exploring every possible way to picture it. As he described in a letter to Ted Welton, "The hope is that a slight modification of one of the pictures will straighten out some of the present troubles. . . . True we only *need* calculate. But a picture is certainly a *convenience* & one is not doing anything wrong in making one up."

To understand the nature of the pictures he was playing with, we first need to explore a little of the remarkable new complexity that Dirac had introduced into quantum mechanics when he discovered his famous equation on the relativistic properties of particles like electrons. As I have described, electrons possess a property called *spin* because they carry with them intrinsic "angular momentum," the property extended objects possess if they are spinning. Classically no such concept exists for a point particle, which cannot behave as if it is spinning because there is no "center" (that is, another point) to spin around. In order to have angular momentum, like a spinning bicycle wheel, for example, classical objects must have spatial extension.

This strange spin angular momentum, which like all things in quantum mechanics is "quantized" (that is, it exists in only integer multiples of some smallest unit), plays a central role in the behavior not only of electrons but in fact of all materials. Electrons orbiting around nuclei in atoms, for example, possess angular momentum, just like planets orbiting the sun do, but the values are quantized, as Neils Bohr was the first to show. It turns out that the internal angular momentum of electrons has a value of one-half of the smallest unit of orbital angular momentum, so we call them spin ½ particles.

This property ultimately explains why solid objects exist and behave the way they do. Wolfgang Pauli, the great Austrian theoretical physicist, had explained that atomic properties could be understood if one postulated what he called an "exclusion principle," as follows: no two electrons, or any other spin ½ particle (protons and neutrons are also such particles) could coexist in exactly the same quantum mechanical state at the same place at the same time.

Two electrons orbiting in a helium atom, for example, could not normally occupy the same orbit. But they can if they are spinning in two different directions so that they are not in identical quantum mechanical states. If we then consider the next lightest element, lithium, for example, which has three electrons orbiting its nucleus, there is no independent option for the third electron, which therefore must orbit the nucleus in a different, presumably higher-energy, orbit. All of chemistry can be understood to result from the application of this simple principle to predict the energy levels of electrons in atoms.

Similarly, if we bring two identical atoms close together, there is not only an electric repulsion between the negatively charged electrons in one atom and those in the other atom, but the Pauli exclusion principle tells us that there is an additional repulsion because no two electrons can be in the same place in the same quantum state. Thus the electrons in one atom are pushed apart from the electrons in the neighboring atoms so they don't overlap in the same position in the same orbital configuration. These two effects resulting from the Pauli exclusion principle combine to determine the mechanical properties of all materials that make up the world of our experience.

The Italian physicist Enrico Fermi next explored the statistics of systems of many identical particles with spin ½ such as electrons and demonstrated that the exclusion principle strongly governed the behavior of these many particle states. We now call all such particles with spin ½, 3⁄2, and so on, *fermions*, after Fermi. Other particles with integral values of spin, including photons—the quanta of electromagnetic fields with spin 1—as well as those particles with no spin whatsoever, are now called *bosons*, after the Indian physicist Satyendra Bose, who, along with Albert Einstein, described the collective behavior of these particles.

BY "PLAYING," AS he later described it, with the mathematical description of spin ½ particles, in 1928 Dirac was able to derive an equation describing electrons that could account for their spin and was in accord with the requirements for how a theory should behave at relativistic velocities according to Einstein's theory of relativity. It was a remarkable achievement, and it had an even more remarkable prediction, so remarkable, in fact, that Dirac and most other leading physicists didn't believe it. The theory predicted that in addition to electrons, there must exist particles just like electrons that were negative energy solutions of the equations. However, since negative energy seemed unphysical—Einstein's equations always associate positive energy with mass—these particles had to be interpreted somewhat differently.

The interpretation that Dirac came up with reminds me of an old joke I once heard about two mathematicians sitting in a bar in Paris looking at a nearby building. Early on in the lunch they see two people walk into the building. During dessert, they observe three people leaving the building. One mathematician then turns to the other and says, "If one more person goes into that building, it will be empty!"

Similarly, if we interpret negative energy as having less

energy than zero, then we might perversely choose to imagine that while an electron has positive energy, and a state with no electrons has zero energy, then a state with negative energy simply has fewer than zero electrons. And a state with a negative energy precisely equal and opposite to the energy of a single electron would thus be described as having one electron less than a state with zero electrons.

While as a formal statement this is consistent, physically it seems ridiculous. What would "one electron less than zero electrons" physically mean? A clue comes from thinking about the charge on an electron: since electrons have negative electric charge, and a state with zero electrons has zero electric charge, then a state with one less than zero electrons would have positive electric charge. Put another way, having a negative number of electrons is equivalent to having a positive number of positively charged particles. Hence, the negative-energy state that appeared in Dirac's equation could be interpreted as representing a positive-energy particle with a charge equal and opposite to the negative charge on the electron.

But there was at least one major problem with this exotic interpretation. *Only one* particle in nature was known to have a positive charge equal and opposite to the charge on the electron: the proton. But the proton does not resemble the electron at all—for example, it is two thousand times heavier.

Earlier, immediately after deriving his equation, Dirac had recognized another important problem with his negative-energy states. Remember that in quantum mechanics all possible configurations are explored as a system evolves. In particular, as he put it, in his new theory "transitions can take place in which the energy of the electron changes from a positive to a negative value even in the absence of any external field, the surplus energy, at least $2mc^2$ in amount, being spontaneously emitted in the form of radiation." Put in simpler language, an electron could spontaneously decay into the positively charged particle corresponding to the negative-energy state. But this would change the total charge of the system, which is not allowed in electromagnetism. Moreover, if the positively charged particle was the much heavier proton, then such a transition would also violate energy conservation.

In order to address these problems, Dirac made a radical proposal. Remember that electrons are fermions, and therefore only one particle can exist in each different quantum state. Dirac imagined what would happen if empty space actually contained an infinite "sea" of negative-energy electrons, and all available quantum states of these particles could therefore be occupied. There would thus be no available states for real positive-energy electrons to decay into. Moreover, he argued that if by some process a negative-energy state became unoccupied, this would leave a "hole" in the distribution. The hole, corresponding to the absence of a negatively charged electron in the sea, could then be identified with a positively charged particle, which he in turn identified with a proton.

The suspension of disbelief involved in Dirac's assertion was enormous. It implied first that the vacuum—that is, empty space—somehow contained an infinite number of unobservable particles corresponding to the filled negative-energy levels, and moreover that the odd hole in these filled levels would be observed as a proton, a particle completely unlike the electron except for the magnitude of its electric charge.

As intellectually courageous as it might have been to propose an infinite sea of negative-energy particles, the proposal that holes in this sea represented protons was a rare act of intellectual cowardice for Dirac. The negative-energy states in his equation appeared completely symmetrical to the positive-energy states, suggesting that they had precisely the *same mass*, in manifest contradiction to the fact that the proton is much heavier. Dirac tried to circumvent this apparent problem by supposing that in the filled sea the interactions between particles would be such that the few holes that might appear would receive additional contributions to their mass from these interactions.

Had he had more courage, Dirac could have simply predicted that these holes represented new elementary particles in nature with a mass equal to that of the electron, but with opposite electric charge. But, as he later said, "I just didn't dare to postulate a new particle at that stage, because the whole climate of opinion at that time was against new particles."

More charitably, perhaps, Dirac might have hoped to explain all of the known elementary particles at that time, electrons and protons, as resulting from different manifestations of a single particle, the electron. This reflects the spirit of physics, to explain manifestly different phenomena as representing merely two different sides of the same coin. Either way, this confusion did not last long. Other well-known physicists who had examined his theory, including Werner Heisenberg, Herman Weyl, and Robert Oppenheimer, correctly inferred that interactions in the "Dirac sea" would never add mass to the holes and result in the holes having a different mass than the electrons. Ultimately even Dirac was forced to recognize that his theory predicted the existence of a new particle in nature, one he called the *anti-electron.*

Dirac had made his concession in 1931, and he did so just in time. It would only take a year for nature to prove him correct, although there was such skepticism about the possibility of new as-of-yet-unobserved elementary particles that even after finding strong evidence of their existence, the first group to observe the anti-electron, or *positron* as it became known, didn't believe its own data.

During the 1930s, before particle accelerators were first developed and built, almost all of the information about elementary particles came from observations of the products of nature's astrophysical accelerators—namely, the cosmic rays that bombard the earth daily, whose origin ranges from as close as our sun, to more energetic sources like exploding stars in distant galaxies at the far reaches of our universe. Two different groups on either side of the Atlantic were examining cosmic ray data in 1932. One group, working in the same laboratory as Dirac at Cambridge in the United Kingdom, under the leadership of Patrick Blackett, told Dirac that they had found evidence of his new particle, but they were too timid to publish their results until they did more tests. In the meantime, perhaps characteristic of a brasher American attitude, Carl Anderson in California published compelling evidence of the existence of the positron in 1933, ultimately leading to the award of a Nobel Prize for his discovery. It is interesting that even after Blackett and his collaborator, Giuseppe Occhialini, were finally induced by Anderson's discovery to publish their results a year later, they were still hesitant to ascribe this particle to Dirac's proposal. Ultimately, by the end of 1933 even these experimenters had to admit that if it walks like a duck, and quacks like a duck, it is probably a duck. The properties Dirac predicted agreed strikingly with observations, and like it or not, it appeared that electrons and positrons—the first example of an antiparticle known in nature—could be created in pairs amid the energetic showers produced by cosmic rays bombarding nuclei.

Suddenly positrons were real. Reflecting on his initial hesitation to accept the conclusions of his theory in predicting the existence of antiparticles, Dirac later said, "My equation was smarter than I was!"

IT WAS IN the context of these exotic and revolutionary developments that Richard Feynman, in 1947 and 1948, set to work to invent new "pictures" to incorporate Dirac's relativistic electrons in his own emerging space-time sum-over-paths picture of quantum mechanics. In doing so, he would find that he needed to reinvent his way of doing physics yet again, even as he tried at the same time to reinvent himself, to sort out the deep emptiness of his personal life.

CHAPTER 8

From Here to Infinity

> It therefore seems that I guessed right, that the difficulties of electrodynamics and the difficulties of the hole theory of Dirac, are independent and one can be solved before the other.
>
> —RICHARD FEYNMAN, IN A LETTER DATED 1947

Perhaps it took a man who was willing to break all of the rules to fully tame a theory like quantum mechanics that breaks all of the rules. As Richard Feynman turned his attention once again to QED, he was already cultivating a reputation for scoffing at society's norms in his job, his love life, and his institutional interactions. Even while at Los Alamos he loved creating havoc—finding holes in security fences, entering through them, and then exiting through the main gate when no record existed of his entering, or picking locks and leaving messages in top-secret safes.

Following Arline's death and his newfound nihilism after the Trinity bomb explosion, he responded to his inner turmoil by lashing out at convention. From then on, he began to revel in being different. While formerly shy with women, he became a womanizer. Within months after Arline's death, while still at Los Alamos, he began dating beautiful women at a frenzied pace. Two years later, when his grief had finally surfaced, he was able to write a letter to Arline, exposing his pain: "I'll bet you are surprised that I don't even have a girl friend (except you sweetheart) after two years. But you can't help it, darling, nor can I—I don't understand it, for I have met many girls and very nice ones . . . but in two or three meetings they all seem ashes."

The liaisons may have left him feeling empty, but they nevertheless continued. When he first arrived at Cornell, he still looked like a student, and in his loneliness he dated undergraduates he met at freshman dances. His pursuit of women was matched in intensity only by his desire to drop them. In 1947, before he provided final grades for his students, he left on a famous cross-country trip with then graduate student Freeman Dyson. The prime purpose of this expedition was to end an entanglement with a woman in Los Alamos. He had continued an intense long-distance courtship with her, and she was causing another woman in Ithaca to lash out at him in jealousy. Meanwhile yet another woman, one of several who apparently had become pregnant by him and aborted their pregnancies, reacted more stoically in a letter to him, in which she also corrected his misspelling of her name.

All the while he remained in Ithaca he never settled into a single abode. He often stayed with friends, usually married ones, and these visits frequently ended badly as a result of his sexual improprieties. A few years later, when he spent a year in Brazil, he actually devised a set of simple rules for seducing women, including prostitutes, at bars. He became famous for seducing women at conferences abroad.

His attractiveness was understandable. He was brilliant, funny, confident, and charismatic in the extreme. He was tall, and had become more handsome as the years passed. His piercing eyes were mesmerizing, and his energy and enthusiasm were addictive.

But it wasn't only in matters sexual that he flaunted convention. Everywhere he encountered what he viewed as nonsense, he rebelled, often independent of standard formalities. An episode with several psychiatrists who performed a second draft physical on Feynman in the summer of 1947 was worthy of an Abbott and Costello skit and later became famous. As a result of his eruptions during the psychiatric interview, he was declared mentally unfit to serve, a conclusion that caused both him and Hans Bethe to erupt in nonstop laughter for half an hour when he returned back to work.

Feynman would later cultivate these sorts of anecdotes as part of the Feynman mythology he liked to encourage. But in 1947 he was not yet famous, and the buildup of his unconventional attitudes and behavior coincided with what became the most intense two-year period of creative activity in his life, a time corresponding with experimental discoveries that made solving an otherwise obscure mathematical problem more urgent if physical progress was to be made.

The experimental discovery of the positron in 1932 provided remarkable vindication of Dirac's relativistic QED, representing the first time in history where the existence of a previously unobserved elementary particle was intuited on the basis of purely theoretical reasoning. However, it added, literally, a frustrating new infinite level of confusion for physicists who were trying to make sense of the predictions of the theory. For once the existence of positrons was verified, the horrible complications introduced by the possibility of a Dirac sea and the interactions of both electrons and now these new particles, positrons, with radiation—the very interactions that Feynman had originally hoped to erase from the quantum theory of electromagnetism—confronted physicists squarely in the face.

While the predicted interactions of single electrons with single photons or even with classical electromagnetic light or radio waves were in remarkable agreement with observations, whenever physicists tried to go beyond this simplest approximation, by including multiple quantum interactions or even attempting to address the long-standing problem of an electron interacting with itself—the very problem that Feynman first tried to tackle in graduate school—their answers remained infinite and thus physically untenable. This inability to make sense of a theory that was clearly correct at some deep level could safely be ignored at the time in almost all practical applications, but it wore on a select group of ambitious physicists like a raw and exposed nerve. A sense of the desperation dominating the scene can be gleaned from statements of several among the great theoretical physicists of the time. Heisenberg wrote in 1929 that he was frustrated trying to understand Dirac's ideas and that he was concerned that he might be "forever irritated by Dirac." Wolfgang Pauli wrote in 1929 about his concerns (which presciently reflected concerns that many physicists, including Feynman, later voiced about more recent developments in physics): "I am not very satisfied. . . . In particular, the self-energy of the electron makes much bigger difficulties than Heisenberg had thought at the beginning. Also the new results to which our theory leads to are very suspect and the risk is very great that the entire affair loses touch with physics and degenerates into pure mathematics."

Heisenberg in turn wrote to Pauli in 1935, "With respect to QED . . . we know that everything is wrong. But in order to find the direction in which we should depart . . . we must know the consequences of the existing formalism much better than we do." He later added in a subsequent paper, "The present theory of the positron and QED must be considered provisional." Even Dirac said of QED in 1937, "Because of its extreme complexity, most physicists will be glad to see the end of it."

The concerns were so great that these physicists, and in particular the great Danish physicist Neils Bohr, worried that perhaps quantum mechanics itself might be at the root of the problem and might have to be replaced by different physics. Bohr wrote to Dirac in 1930: "I have been thinking a good deal of the relativity problems lately and believe firmly that the solution of the present troubles will not be reached without a revision of our general physical ideas still deeper than that contemplated in the present quantum mechanics."

Even Pauli suggested, in 1936, that quantum mechanics might have to be revised when dealing with systems, like Dirac's hole theory, that allow an infinite number of particles to be present in empty space. For beyond the well-known infinity associated with the self-energy of the electron due to its interaction with its own electromagnetic field, Dirac's introduction of antiparticles created another new class of infinite interactions that further muddied the quantum waters. These new interactions involved, not the intermediate photons that Feynman and Wheeler had worked so hard to get rid of, but pairs of "virtual" electrons and positrons.

Since physicists now knew that particles and antiparticles can annihilate into pure radiation, and that the reverse process—the complete conversion of energy into mass, could also, in principle, occur. There are constraints on this conversion, however. For example, an electron and its antiparticle, a positron, cannot annihilate into a single particle of radiation (a photon) for the same reason that when a bomb explodes, all the pieces do not fly off in a single direction. If the electron and positron come at each other with equal and opposite velocities, then their total momentum is zero. If a single photon were produced in their annihilation, it would fly off at the speed of light in some direction, carrying nonzero momentum. Thus, at least two photons must be produced when an electron and positron annihilate, so the two emitted particles can fly off in equal and opposite directions as well. Similarly, a single photon cannot suddenly convert into a positron and electron pair. Two photons must come together to produce the final pair.

But remember that with virtual particles all bets are off, and energy and momentum need not be conserved as long as the virtual particle disappears in a time sufficiently short so that it cannot be measured directly. Thus, a virtual photon can spontaneously transform into an electron-positron pair, as long as the electron-positron pair then annihilates and transforms back into the single virtual photon on a short timescale.

The process involving a photon momentarily splitting up into an electron-positron pair is called *vacuum polarization.* It gets this name because in a real medium such as any solid object made of atoms, which contains both positive and negative charges, if we turn on a large external electric field, we can "polarize" the medium by separating charges of different types—negative charges will be pushed in one direction by the field, while positive charges will be pulled in the other direction. Thus a neutral material will remain neutral, but the charges of different signs will spatially separate. This is what momentarily happens in empty space as a photon splits temporarily into a negatively charged electron and its positively charged antiparticle, a positron. Thus, empty space gets briefly polarized.

Whatever we call it, an electron, which previously had to be thought of as having a cloud of virtual photons around it, now had to be thought of as being surrounded by a cloud of virtual photons *plus* electron-positron pairs. In some sense, this picture is just another way of thinking about Dirac's interpretation of positrons as "holes" in an infinite sea of electrons in the vacuum. Either way, once we include relativity, and the existence of positrons, the theory of a single electron turns into a theory of an infinite number of electrons and positrons.

Moreover, just as emission and absorption of virtual photons by a single electron produced an infinite electron self-energy in calculations, the production of virtual particle-antiparticle pairs produced a new infinite correction in QED calculations. Recall once again that the electric force between particles can be thought of as being due to the interchange of virtual photons between those particles. If the photons could now split up into electron-positron pairs, this process would change the strength of the interaction between particles and would thus shift the calculated energy of interaction between an electron and a proton in an atom like hydrogen. The problem was, the calculated shift was infinite.

The frustrating fact was that the Dirac theory produced very accurate predictions for the energy levels of electrons in atoms as long as only the exchange of single photons was considered and the annoying higher-order effects that produced infinities were not. Morever Dirac's prediction of positrons had been vindicated by experimental data. Were it not for these facts, many physicists, as Dirac implied, would have preferred to simply dispense with QED altogether.

IT TURNED OUT that what was needed to resolve all of these difficulties was neither a wholesale disposal of quantum mechanics, nor dispensing with all these virtual particles, but rather developing a deeper understanding of how to implement the basic principles of quantum theory in the context of relativity. It would take a long circuitous route, and the guidance brought by key experiments, before this fact, hidden in a mire of crushingly complex calculations, would become clear, both to Feynman and to the rest of the world.

The process of discovery began slowly and confusedly, as it usually does. After his *Reviews of Modern Physics* paper was completed, Feynman turned his attention once again to Dirac's theory. He had decided physics was fun again, and in spite of his otherwise unsettled personal situation, his focus never strayed far from the problem that had obsessed him since he had been an undergraduate—the infinite self-energy of an electron. It was a puzzle he hadn't yet solved, and it was contrary to his nature to let it go.

He began with a warm-up problem. Since the spin of the electron makes sense only quantum mechanically, Feynman began by trying to understand whether he could account for spin directly within his sum-over-paths formalism. One complication of Dirac's theory is that a single equation had four separate pieces: one to describe spin-up electrons, one to describe spin-down electrons, one to describe spin-up positrons, and one to describe spin-down positrons. Since the normal conception of spin requires three dimensions (a two-dimensional plane to spin in and a perpendicular axis to spin around), Feynman reasoned that he could make the problem simpler if he first tried to consider a world with just one spatial dimension and one time dimension, where the different sorts of allowed paths were also trivial. Paths would just involve travel back and forth in the one space dimension, namely, a line.

He was able to derive the simplified version of the Dirac equation appropriate for such a two-dimensional world if every time an electron "turned around" from rightward motion to leftward motion, the probability amplitude for that path got multiplied by a "phase factor," which in this case was a "complex number," an exotic number that involved the square root of –1. Complex numbers can appear in probability amplitudes, and actual probabilities depend on the square of these numbers, so that only real numbers appear in the final result.

The notion that somehow spin might produce additional phases when one is calculating probability amplitudes was prescient. However, when Feynman tried to move beyond one spatial dimension and associate more complicated phase factors as electrons turned corners and went off at different angles, he got nonsensical answers and couldn't get results that corresponded to Dirac's theory.

Feynman kept trying different, diffuse alternatives to reformulate the theory, but he made little progress. However, there is one area where his sum-over-paths methodology was particularly useful. Special relativity tells us that one person's "now" may not be another person's "now"—namely, observers in relative motion have different notions of simultaneity. Special relativity explains how this local notion of simultaneity is myopic, and how the underlying physical laws are independent of different observers' individual preferences for "now."

The problem with the conventional picture of quantum mechanics was that it depended explicitly on defining a "now," in which an initial quantum configuration was established, and then determining how this configuration would evolve to a later time. In the process, the relativistic invariance of physical laws gets buried, because the minute we choose a particular spatial frame to define the initial wave function and an instant of time to call $t = 0$, we lose explicit contact with the underlying relativistic, frame-independent beauty of the theory.

Feynman's space-time picture, however, was precisely tuned to make the relativistic invariance of the theory manifest. In the first place, it was defined in terms of quantities—Lagrangians—which can be written in an explicit relativistically invariant form. And secondly, since the sum-over-paths approach inevitably deals with all of space and time together, we do not have to restrict ourselves to defining specific instants in time or space. Thus, Feynman had trained himself to combine quantities in QED—that otherwise might be considered separately—together into combinations that behaved in a way in which the properties of relativity remained manifest. While he had made no real headway in explicitly reformulating the Dirac theory from first principles in any way that resolved the issues he was concerned with, the tricks he had developed would prove of crucial importance later in the ultimate solution.

THE SOLUTION CAME into view, as it usually does, with an experiment. Indeed, while theorists normally take their guidance from experimental results, it is hard to overstress how literally important they were to driving progress in this case. Up to this point, the infinities were frustrating to theorists, but that is about all. As long as the zero-order predictions of Dirac's equation were sufficient to explain, within the achievable experimental accuracy, all the results of atomic physics, theorists could worry about the fact that higher-order corrections, which should have been small, were in fact infinite, but the infinities were not yet a real practical impediment to using the theory in a physical context.

Theorists love to speculate, but I have found that until experimentalists actually produce concrete results that probe a theory at a new level, it is hard for theorists to take even their own ideas seriously enough to rigorously explore all of their ramifications, or to come up with practical solutions to existing problems. The preeminent U.S. experimental physicist at the time, I. I. Rabi, who had made Columbia University the experimental capital of the world for atomic physicists, mused on this inability of theorists to rise to the challenges of QED in the absence of experimental guidance. In the spring of 1947 he is reported to have said to a colleague over lunch, "The last eighteen years have been the most sterile of the century."

All that changed within a few months. As I have already described, until that time the lowest-order calculations performed with Dirac's relativistic theory produced results for the spectrum of energy levels of electrons bound to protons in hydrogen atoms which were sufficient not only to understand the general features of the spectrum, but also to strike quantitative agreement with observation, aside from a few possible inconsistencies that emerged at the very limit of experimental sensitivities and were thus largely ignored. That was, however, until a courageous attempt by the American physicist Willis Lamb—who worked in Rabi's Columbia group, and who was one of the last of a breed of physicists who were equally adept in the laboratory and performing calculations—changed everything.

Recall that the initial great success of quantum mechanics in the early decades of the twentieth century lay in explaining the spectrum of light emitted by hydrogen. Neils Bohr was the first to propose a rather ad hoc quantum mechanical explanation for the energy levels in hydrogen: that electrons were able to jump between only fixed levels as they absorbed or emitted radiation. Later, Schrödinger, with his famous wave equation, showed that the electron energy levels of hydrogen could be derived precisely from using his "wave mechanics," instead of by fiat as in the case of the Bohr atom.

Once Dirac derived his relativistic version of QED, physicists could attempt to replace the Schrödinger equation with the Dirac equation in order to predict energy levels. They did this and discovered that the energy levels of different states were "split" by small amounts, owing to relativistic effects (for example, more energetic electrons in atoms would be more massive, according to relativity) and to the nonzero spin of electrons incorporated into Dirac's equation. Lo and behold, the predictions of Dirac agreed with observations of the more finely resolved spectra from hydrogen, where what were otherwise seen as single frequencies of emission and absorption were now shown to be split into two different, very finely separated frequencies of light. This *fine structure* of the spectrum, as it became known, was yet another vindication of the Dirac theory.

In 1946 Willis Lamb decided to measure the fine structure of hydrogen more accurately than it had ever been measured before, in order to test the Dirac theory. His proposal for this experiment explained his motivation: "The hydrogen atom is the simplest one in existence, and the only one for which essential exact theoretical calculations can be made. . . . Nevertheless, the experimental situation at present is such that the observed spectrum of the hydrogen atom does not provide a very critical test. . . . A critical test would be obtained from a measurement of . . . fine structure."

On April 26, 1947 (back in the days when the time between proposing an experiment and completing it in particle physics was on the order of months, not decades),

Lamb and his student Robert Retherford successfully completed a remarkable measurement that had previously been unthinkable. The result was equally astounding.

The lowest-order Dirac theory, like the Schrödinger theory before it, predicted that the same energy would be ascribed to two different states with the same total angular momentum of the electron in hydrogen—arising from the sum of the spin angular momentum and the orbital angular momentum—even if the separate pieces of the sum were different in the two states. However, Lamb's experiment conclusively proved that the energy of electrons in one state differed from that of electrons in the other. Specifically he observed that the transitions of electrons between one state and a fixed higher state in hydrogen resulted in the emission or absorption of light whose frequency differed by about a billion cycles per second compared to transitions of electrons between the other state and the fixed higher state. This may seem like a lot, but the characteristic frequencies of light emitted and absorbed between energy levels in hydrogen were about ten million times bigger than even this frequency difference. Lamb was therefore required to measure frequencies with an accuracy of better than one part in ten million.

The state of theoretical physics following Dirac's coup was such that the impact of this almost imperceptibly small, yet clearly nonzero, difference with the predictions of Dirac's theory was profound. Suddenly, the problem with Dirac's theory was concrete. It did not revolve around some obscure and ill-defined set of infinite results, but it now came down to real and finite experimental data that could be computed. Feynman later described the impact in his typically colorful fashion: "Thinking I understand geometry, and wanting to fit the diagonal of five-foot square I try to figure out how long it must be. Not being very expert I get infinity—useless. . . . It is not philosophy we are after, but the behavior of real things. So in despair, I measure it directly—lo, it is near to seven feet—neither infinity, nor zero. So, we have measured these things for which our theory gives such absurd answers."

In June of 1947 the National Academy of Sciences convened a small conference of the greatest theoretical minds working on the quantum theory of electrodynamics (fortunately Feynman's former supervisor, John Wheeler, was one of the organizers so Feynman was invited) in a small inn on Shelter Island, off Long Island, New York. The purpose of the "Conference on the Foundations of Quantum Theory" was to explore the outstanding problems in quantum theory that had been set aside during the war, when Feynman and his colleagues were laboring on producing the atomic bomb. In addition to Feynman, all the leading lights from Los Alamos were there, from Bethe to Oppenheimer, and the young theoretical superstar Julian Schwinger.

It was at this small meeting, which began in suitably dramatic fashion, with the police escorting the famous war-hero "atomic" scientists through Long Island, that Lamb presented the results of his experiment. This was the highlight of the meeting, which Feynman later referred to as the most important conference he had ever attended.

As far as Feynman's work was concerned, however, and probably for all theorists thinking about the problems of QED, the most important outcome of the conference was not a calculation that Feynman performed, but rather a calculation that his mentor, Hans Bethe, performed on the train trip back to Ithaca from New York City, where Bethe had stayed for a few days to visit his mother. Bethe was so excited by the result he had obtained that he phoned Feynman from Schenectady to tell him the result: In his typical fashion, when finally presented with an experimental number, Bethe found it irresistible to use whatever theoretical machinery was at his disposal, no matter how limited, to derive a quantitative prediction to be compared with the experimental result. To his immense surprise and satisfaction, even without a full understanding of how to deal with the strange infinities of QED, Bethe claimed to understand the magnitude and origin of the frequency shift that had already become known as the *Lamb shift.*

For Feynman, Schwinger, and the rest of the community, the gauntlet had been laid.

CHAPTER 9

Splitting an Atom

A very great deal more truth can
become known than can be proven.

—RICHARD FEYNMAN,
NOBEL LECTURE, 1965

When Willis Lamb presented his result to begin the Shelter Island conference, the question immediately arose as to what could have caused the discrepancy between observations and Dirac's QED theory. Oppenheimer, who dominated the meeting, suggested that perhaps the source of the frequency shift might be QED itself, if anyone could actually figure out how to tame the unphysically infinite higher-order corrections in the theory. Bethe's effort to do just that built on ideas from Oppenheimer and the physicists H. A. Kramers and Victor Weisskopf, who later would take a leave from MIT to become the first director of the European Laboratory for Nuclear Research, called CERN, in Geneva.

Kramers emphasized that since the problem of infinite contributions in electromagnetism went all the way back to the classical self-energy of an electron, physicists should focus on observable quantities, which were of course finite, when expressing the results of calculations. For example, the electron mass term that appeared in the equations, and in turn received infinite self-energy corrections, should not be considered to represent the measured physical mass of the particle. Instead, call this the *bare mass.* If the bare mass term in the equation was infinite, perhaps the sum of this term and the infinite self-energy correction could be made to cancel, leaving a finite residue that could be equal to the experimentally measured mass.

Kramers proposed that all of the infinite quantities that one calculated in electrodynamics, at least for electrons moving nonrelativistically, could be expressed in terms of the infinite self-energy contribution to the electron rest mass. In this case as long as one removed this single infinite quantity by expressing all results in terms of the finite measured mass, then all calculations might yield finite answers. In doing so, one would change the magnitude, or the *normalization*, of the mass term appearing in the fundamental equations, and this process became known as *renormalization*.

Weisskopf and Schwinger explicitly considered the relativistic quantum theory of electrodynamics in an effort to implement this idea. In particular, they demonstrated that the infinity one calculated in the self-energy of the electron actually became somewhat less severe when one incorporated relativistic effects.

Motivated by these arguments, Bethe performed an approximate calculation of such a finite contribution. As Feynman later put it in his Nobel address, "Prof. Bethe . . . is a man who has this characteristic: If there's a good experimental number you've got to figure it out from theory. So, he forced the quantum electrodynamics of the day to give him an answer to the separation of these two levels [in hydrogen]." Bethe's reasoning was undoubtedly something like this: if the effects of electrons and holes and relativity seemed to tame infinities somewhat, then perhaps one could do a calculation with the nonrelativistic theory, which was much easier to handle, and then simply ignore the contribution of all virtual photons higher than an energy equal to approximately the measured rest mass of the electron. Only when total energies exceed this rest mass do relativistic effects kick in, and perhaps when they do, they ensure that the contribution from virtual particles of higher energy become irrelevant.

When Bethe performed the calculation with this arbitrary cutoff in the energy of virtual particles, the predicted frequency shift between light emitted or absorbed by the two different orbital states in hydrogen was about 1,040 megacycles per second, which was in very good agreement with the observation of Lamb.

Feynman, as brilliant as he was, later remembered not fully appreciating Bethe's result at the time. It was only later, at Cornell, when Bethe gave a lecture on the subject suggesting that if one had a fully relativistic way of handling the higher-order contributions in the theory one might be able to not only get a more accurate answer but also demonstrate the consistency of the ad hoc procedure he had employed, that Feynman understood both the significance of Bethe's result and how all the work he had done up to that point could allow him to improve on Bethe's estimate.

Feynman went up to Bethe after his lecture and told him, "I can do that for you. I'll bring it in for you tomorrow." His confidence was based on his years of labor reformulating quantum mechanics using his action principle and the sum over paths, which provided him with a relativistic starting point that he could use in his calculations. The formalism he had developed in fact allowed him to adjust the possible paths of particles in a way that would constrain the otherwise infinite terms in the quantum calculation by effectively limiting the maximum energy of the virtual particles that enter into the calculation, but did so in a way that was consistent also with relativity, just as Bethe had requested.

The only problem was that Feynman had never actually worked through a calculation of the self-energy of the electron in the quantum theory, so he went to Bethe's office, where Bethe could explain to him how to do the calculation, and Feynman in turn could explain to Bethe how to use his formalism. In one of those serendipitous accidents that affect the future of physics, when Feynman went to visit Bethe and they worked their calculations out at the blackboard, they made a mistake. As a result, the answer they got at the time was not only not finite, but the infinities were actually worse than had appeared in the nonrelativistic calculation, making it harder to isolate what were the finite pieces.

Feynman went back to his room, certain that the correct calculation should be finite. Ultimately, in a typical Feynmanesque way, he decided he had to teach himself in excruciating detail how to do the self-energy calculation in the traditional complicated way using holes, negative-energy states, and so on. Once he knew in detail how to do the calculation the traditional way, he was confident he would be able to repeat it using his new path-integral formalism, doing the modifications necessary to make the result finite but in a way where relativity remained manifestly obeyed.

When the dust had settled, the result was just what he had hoped for. Expressing everything in terms of the experimentally measured rest mass of the electron, Feynman was able to get finite results, including a highly accurate result for the Lamb shift.

As it turned out, others had also been able to do a relativistic calculation at around the same time, including Weisskopf and his student Anthony French, as well as Schwinger. Schwinger, moreover, was able to show that taming precisely the same infinities that resulted in a finite and calculable Lamb shift also allowed a calculation of another experimental deviation from the predictions of the uncorrected Dirac theory discovered by Rabi's group at Columbia. This effect had to do with the measured magnetic moment of the electron.

Since the electron acts like it is spinning, and since it is charged, electromagnetism tells us that it should also behave like a tiny magnet. The strength of its magnetic field should therefore be related to the magnitude of the electron's spin. But measurement of the strength revealed that it deviated from the simple lowest-order prediction by about 1 percent. This is a small amount, but nevertheless the accuracy of the measurement was such that the difference from the prediction was real and significant. One therefore needed to understand the theory to a higher order to know if it agreed with experimental data.

Schwinger showed that the same type of calculation, isolating the otherwise infinite pieces and modifying them in a well-defined way, and then expressing all of the calculated results in terms of quantities like the measured rest mass of the electron, produced a predicted shift in the magnetic moment of the electron that conformed with the experimental result.

Rabi wrote Bethe an elated note on hearing about Schwinger's calculation, and Bethe replied, referring back to Rabi's experiments, "It is certainly wonderful how these experiments of yours have given a completely new slant to a theory and the theory has blossomed out in a relatively short time. It is as exciting as in the early days of quantum mechanics."

QED had at last begun to emerge from a long murky initiation. In the years since that time, the predictions of the theory have agreed with the results of experiments to an accuracy that is unparalleled anywhere else in all of science. There is simply no better scientific theory in nature from this point of view.

IF FEYNMAN HAD been only one of a crowd that had correctly shown how to calculate the Lamb shift we probably would not be memorializing his contributions today. But the real value of his efforts at calculating the Lamb shift, and his understanding of how to tame the infinities involved in the calculation, was that he began to calculate more and more things. And in the process, he used his formidable mathematical skills, along with the intuition he had developed in the process of reformulating quantum mechanics, to gradually develop a whole new way of picturing the phenomena involved in QED. And he did so in a way that produced a remarkable new way of calculating with the theory, based on diagrammatic space-time pictures, which were themselves founded on a sum-over-paths approach.

FEYNMAN'S APPROACH TO resolving the problems of QED was both highly original and highly scattershot. He often simply guessed what the likely formulas should be and then compared his guesses in different contexts with available known results. Furthermore, while his space-time approach allowed him to write down mathematics in a manner that was in accord with relativity, the actual calculations he performed were not derived directly from any systematic mathematical framework in which relativity and quantum mechanics were unified—even if after the fact everything worked out correctly.

A systematic framework for combining relativity and quantum mechanics had in fact existed since at least the 1930s. It was called quantum field theory, and it was intrinsically a theory of infinitely many particles, which is why Feynman probably steered away from it. In classical electromagnetism, the electromagnetic field is a quantity that is described at every point in space and time. When treating the field quantum mechanically, one finds that it can be thought of in terms of elementary particles—in this case, photons. In quantum field theory, the field can be thought of as a quantum object, with a certain probability of creating (or destroying) a photon at each point in space. This allows the existence of a possibly infinite number of virtual photons to be temporarily produced by fluctuations in the electromagnetic field. It was precisely this complication that motivated Feynman to originally reformulate electrodynamics in a way in which these photons disappeared completely and in which there were direct interactions of charged particles. These direct interactions he then handled quantum mechanically using his sum-over-paths approach.

In the midst of his tinkering with how to incorporate Dirac's relativistic theory of electrons into his calculational framework for QED, Feynman stumbled upon a beautiful mathematical trick that simplified tremendously the calculations and did away with the need to think of particles and "holes" as separate entities. But at the same time, this trick makes manifest the fact that the moment relativity is incorporated into quantum mechanics, one can no longer live in a world where the number of possible particles is finite. Relativity and quantum mechanics simply *require* a theory that can handle a possibly infinite number of virtual particles existing at any instant.

The trick Feynman used hearkened back to the old idea that John Wheeler proposed to him one day when he argued that all electrons in the world could be thought of as arising from a single electron, as long as that electron was allowed to go backward as well as forward in time. An electron going backward in time would appear just like a positron going forward in time. In this way, a single electron going forward and backward in time (and masquerading as a positron when it was doing the latter) could reproduce itself a huge number of times at any instant. Naturally, Feynman pointed out the logical flaw in this picture when he argued that if it were true, there would be as many positrons as electrons around at any instant and there aren't.

Nevertheless, the idea that a positron could be thought of as an electron traveling backward in time was an idea— Feynman suddenly realized all those years later—that he could exploit in a different context. When trying to do relativistic calculations, where both electrons and holes normally had to be incorporated, he recognized that he could get the same results by including just electrons in his space-time picture, but allowing processes where the electrons went both forward and backward in time. (The idea that my high school physics teacher mangled a bit when he tried to get me more interested in physics that summer afternoon long ago.)

To understand how Feynman's unified treatment of positrons and electrons arose, it is easiest to begin to think in terms of the diagrams that Feynman eventually began to draw for himself to depict the space-time processes that arose in his sum-over-paths view of quantum mechanics.

Consider a diagram describing the space-time process of two electrons exchanging a virtual photon, emitted at *A* and absorbed at *B*:

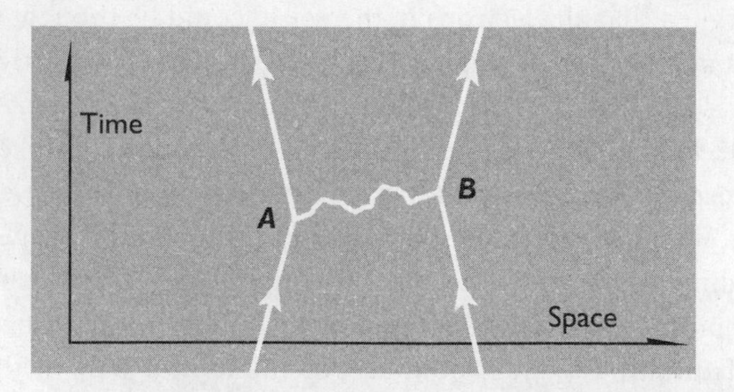

In order to calculate the quantum mechanical amplitude for such a process, we would have to consider all possible space-time paths corresponding to the exchange of a virtual photon between the two particles. Since we don't observe the photon, the following process, in which the photon is emitted at *B* and absorbed at *A*, when *B* is earlier than *A*, also contributes to the final sum:

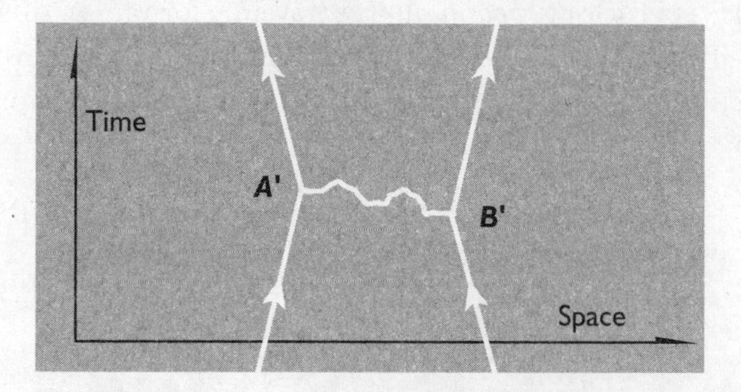

Now, there is another way of thinking about the two separate diagrams. Remember that we are dealing with quantum mechanics, and therefore in the time between measurements, anything that is consistent with the Heisenberg uncertainty principle is allowed. Thus, for example, the virtual photon is not restricted to travel at exactly the speed of light for the entire time it travels between the two particles. But special relativity says that if it is traveling faster than light, then in some frame of reference it would appear to be going backward in time. If it is going backward in time, then it can be emitted at *A* and absorbed at *B*. In other words, the second diagram corresponds to a process identical to the first, except that in the latter case the virtual photon is traveling faster than light.

In fact, while Feynman never explicitly described it at that time, as far as I know, this same effect explains why a relativistic theory of electrons—that is, Dirac's theory—*requires* antiparticles. A photon, which doesn't have any electric charge, traveling backward in time from *A* to *B*, just looks like a photon traveling forward in time from *B* to *A*. But a charged particle traveling backward in time looks like a particle of *opposite* charge traveling forward in time.

Thus, the simple process where an electron (e–) travels between two points, pictured in this space-time diagram:

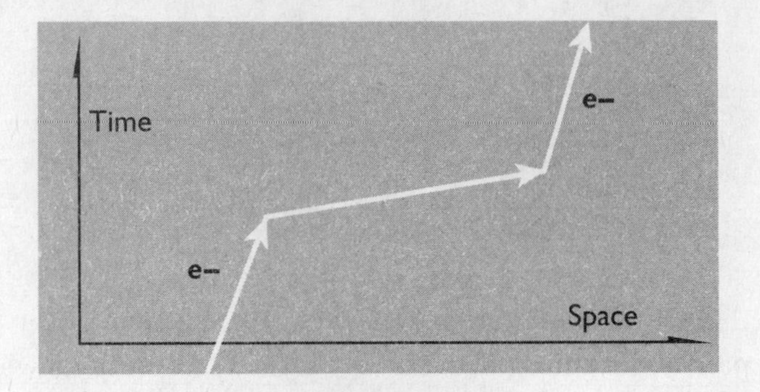

must also be accompanied by this process:

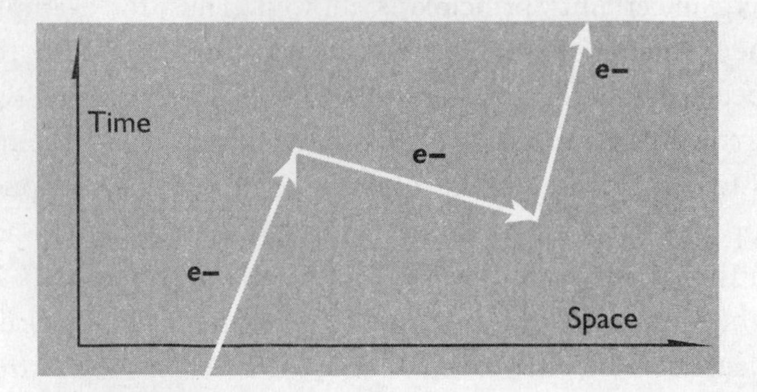

But the latter process could also be described with an intermediate positron (e+), as follows:

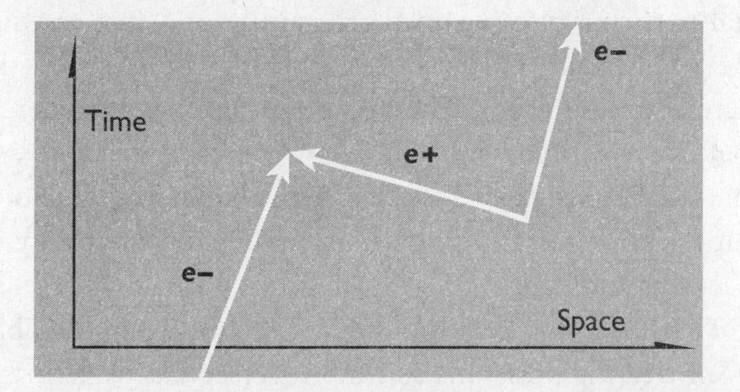

In other words, it appears that a single electron begins its journey, and at another point an electron-positron pair is created from empty space, and a virtual positron travels forward in time, ultimately annihilating with the first electron, leaving only the single final electron at the end of the journey.

Feynman later beautifully described this situation in his 1949 paper "The Theory of Positrons." His famous analogy was a bombardier looking down at a road from his scope on an airplane (the recent world war had undoubtedly influenced Feynman's choice of analogies here): "It is as though a bombardier watching a single road through the bombsight of a low flying plane suddenly sees three roads and it is only when two of them come together and disappear again that he realizes that he has simply passed over a long switchback in a single road."

Thus particle number in a relativistic quantum theory has to be indeterminate. Just when we think we have one particle, a particle-antiparticle pair can pop out of the vacuum, making it three. After the antiparticle annihilates one of the particles (either its partner or the original particle), once again there is only one, just as what the bombardier would see through his bomb-sight if he were counting roads. The key point again is not merely that this is *possible* but that it is *required* by relativity, so that with hindsight we see that Dirac had no other choice but to introduce antiparticles in his relativistic theory of electrons and light.

That Feynman was the one to point out the possibility of treating positrons as time-reversed electrons in his diagrams is fascinating because it immediately implies that his earlier aversion to quantum field theory was misplaced. His diagrammatic space-time expansion for calculating physical effects in QED implicitly contained within it the physical content of a theory where particles could be created and destroyed and particle number during intermediate steps of a physical process was therefore indeterminate. Feynman had been forced by physics to reproduce the physical content of quantum field theory. (In fact, in an obscure 1941 paper that predated Feynman's by eight years, the German physicist Ernst Stückelberg had independently been driven to consider space-time diagrams, and positrons as time-reversed electrons, although he was not sufficiently driven to carry through the program that Feynman ultimately carried out with these tools.)

Now that we are familiar with the diagrams that would eventually become known as *Feynman diagrams*, we can depict the events that correspond to the otherwise infinite processes associated with the electron self-energy and vacuum polarization:

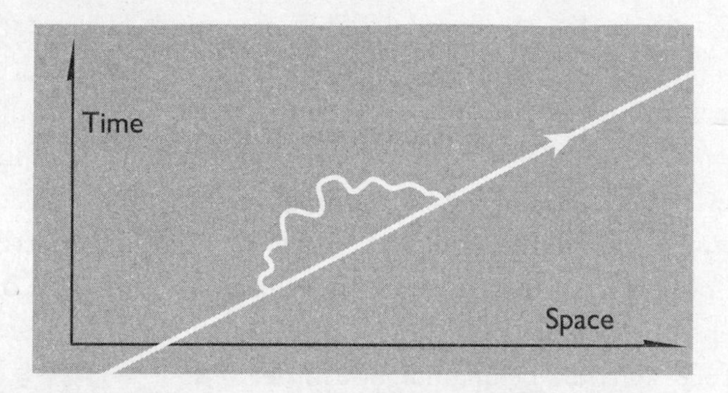

Self-energy (an electron interacting with its own electromagnetic field)

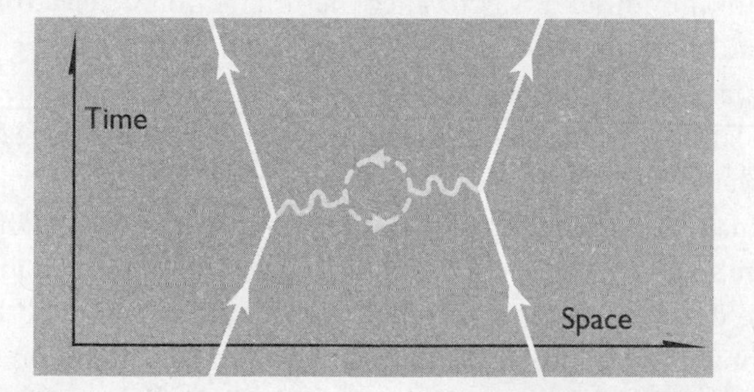

Vacuum polarization (splitting of a virtual photon into electron-positron pair)

For Feynman there was a fundamental difference between these two diagrams, however. The first diagram he could imagine occurring naturally as an electron emitted a photon and later reabsorbed it. But the second diagram seemed unnatural because it would not result from the trajectory of a single electron moving and interacting backward and forward in space and time, and he felt that such trajectories were the only appropriate ones to incorporate in his calculations. As a result, he was wary of the need to include these new processes, and did not originally do so. This decision caused Feynman a number of problems as he tried to derive a framework in which all of the infinities of Dirac's theory could be obviated, and in which predictions for physical processes could be unambiguously derived.

The first great success of Feynman's methods involved calculating the self-energy of the electron. Most important, he found a way to alter the interaction of electrons and photons at very small scales and very high energies in a manner that was consistent with the requirements of relativity. Pictorially this results from considering the case where the loop in the self-energy diagram becomes very small, and then altering the interactions for all loops that are small and smaller. In this way a provisional result could be derived, which is finite. Moreover, this result could be shown to be independent of the form of the alteration of the interactions for small loops in the limit that the loops become smaller and smaller. Most important, as I stressed earlier, because the loops take into account an arbitrary time of emission and absorption and at the same time include objects going forward and backward in time, the form of his alteration did not spoil the relativistic behavior of the theory, which should not depend on any one observer's definition of time.

As Kramers and others had predicted, the key was making these altered-loop contributions finite, or *regularizing* them, as it was later called, in a way that was consistent with relativity. Then if one expressed the corrections to physical quantities, such as the energy of an electron in the field of a hydrogen atom, in terms of the physical mass and physical charge of the electron, the remainder, after canceling out the term that would otherwise become infinite without the alteration for small loops, was both finite and independent of the explicit form of the alteration one made. More important, it remained finite even if the size scale at which one alters the loop diagrams is decreased to zero, where the loop diagram would otherwise become infinite. Renormalization worked. The finite correction agreed reasonably well with the measured Lamb shift, and electrodynamics as a quantum theory was vindicated.

Unfortunately, however, the same type of procedure that Feynman used to change the theory on small distance scales when considering the self-energy of the electron did *not* work when considering the infinite impact of vacuum polarization diagrams. Feynman could find no alteration for small electron-positron loops that maintained the nice mathematical properties of the theory without such an alteration. This is probably another reason, in addition to his sense that these diagrams might not be physically appropriate to the problem at hand, why he ignored them in his original Lamb shift calculation.

Feynman wrestled with this problem on and off during 1948 and 1949. He was able to derive quantitatively accurate results by intuitively supposing that various extra terms in his equations, induced by altering the form of the electron-positron loop, were likely to be unphysical, because they did not respect the mathematical niceties of QED, and therefore could be safely ignored. This unsatisfactory situation was resolved when late in 1949 Hans Bethe informed Feynman of a trick Wolfgang Pauli discovered that allowed a mathematically consistent alteration to be introduced in vacuum polarization diagrams.

Once Feynman incorporated this scheme in his calculations, all of the infinities of QED could be tamed, and every physical quantity could be calculated to arbitrarily high precision and compared with experimental results. From the practical perspective of having a theory of electrons and photons with which one could calculate finite (and accurate) predictions for all processes, Feynman had achieved the goal that had first motivated him in his graduate work with Wheeler.

But Feynman didn't easily let go of his intuitive doubts, and in his famous 1949 paper, "A Space-Time Approach to QED," which outlined his diagrammatic techniques and results, he added a footnote suggesting that one needed Lamb shift experiments that were more precise, to see if the small contribution that he and others had by then calculated to be due to vacuum polarization effects was real.

It was.

CHAPTER 10

Through a Glass Darkly

My machines came from too far away.

—RICHARD FEYNMAN, TO
SYLVAN SCHWEBER, 1984

Feynman's tentative reaction to his own results may seem surprising, but it is not. Nor is it unique. The right answers in science are not always obvious at the time they are developed. When working tentatively, at the edge of knowledge, with many wrong twists, blind alleys, and dead ends, it is easy to be skeptical when nature seems to obey the mathematics envisaged at one's desk. Hence, this is not the end of this part of Feynman's story, at least if we are to understand his true scientific legacy. Instead, we need to examine some associated events, personalities, and twists of fate that actually govern history, the kind of things that one often avoids when attempting to present an after-the-fact logical exposition of scientific concepts.

Two personalities dominated the immediate environment in which Feynman's discoveries were developed and perceived: Julian Schwinger and Freeman Dyson. We have already encountered Julian Schwinger, the boy wonder. Like Feynman, he was drawn to the most important fundamental question in theoretical physics at the time: how to turn

QED into a consistent theory of nature. Like Feynman, Schwinger had contributed to the war effort, and his work then would also have a profound influence on his approach. Schwinger, who worked at the MIT Radiation Laboratory, began to follow an engineering approach based on classical electrodynamics, focusing on sources and responses. And like Feynman he was fiercely driven by a competition primarily with himself and what he thought he should be able to do.

But the similarities stop there. While also heralding from New York, Schwinger grew up in Manhattan, a world away from Long Island, and perhaps no physicist could have projected a more different aura. Brilliant and polished, he was recruited to Columbia by Rabi at age seventeen after he helped resolve a debate Rabi was having with a colleague in the hallway about a subtle point of quantum mechanics. At age twenty-one, he received his PhD, and eight years later became the youngest tenured professor in Harvard's history. He projected an air of supreme confidence and organization. Even though Schwinger always lectured without notes, everything seemed planned, from the place where the chalk would first hit the blackboard to the place where it would last leave it, or rather the places, as he sometimes wrote with both hands. The flow of ideas might be complicated, many would say more complicated than it needed to be, but it was precise, logical, and nothing if not elegant.

Feynman's brilliance manifested itself instead in a kind of intellectual impatience. If he was interested in solving something he would thrust forward to get to the answer and then work backward to understand and fill in the steps. At times he had little patience for those who couldn't keep up, and few could. As he said, in responding to a later job offer from Caltech, "I do not like to suggest a problem and suggest a method for its solution and feel responsible after the student is unable to work out the problem by the suggested method by the time his wife is going to have a baby so that he cannot get a job. What happens is that I find that I do not suggest what I do not know will work and the only way I know it works is by having tried it at home previously, so I find the old saying that 'A Ph.D. thesis is research done by a professor under particularly trying circumstances' is for me the dead truth."

For this reason, perhaps, Feynman had very few successful students in physics. Schwinger, on the other hand, advised over 150 doctoral students during his career, three of whom later won the Nobel Prize, two in physics and one in biology. It is therefore not surprising that the physics community flocked to listen to Schwinger, not least when he was trying to solve the greatest outstanding puzzle at the forefront of fundamental physics.

Following the conference at Shelter Island, Schwinger also threw himself at the task of deriving a relativistically consistent calculation in QED of one of the quantities that differed from the lowest-order predictions of Dirac's theory, the anomalous magnetic moment of the electron. Using a set of tools he developed to address the problems of QED, and the renormalization ideas that he, Victor Weisskopf, and H. A. Kramers had been promoting, he was the first to arrive at an answer, late in 1947.

Almost immediately word began to spread in the physics community of Schwinger's achievement. At the January 1948 meeting of the American Physical Society—the big physics meeting of the year—Schwinger gave an invited lecture titled "Recent Developments in QED." The interest was so great that he was asked to repeat the lecture later that day, and then again to an even larger audience, to satisfy the demand to hear his results.

Meanwhile, Feynman got up after the talk and reported that he had also computed the same quantities that Schwinger had, and in fact claimed some more generality in the anomalous magnetic moment calculation. But he had yet to explain his methods, so his announcement had significantly less impact.

Feynman's ideas generated less interest in the scientific community at the time, not because he did not have the same forum at that meeting, but because his entire approach to addressing the problems he eventually solved was unique. He had long ignored the safety net of conventional quantum field theory, and while his diagrams allowed him to calculate remarkable results, to others they may have looked like doodles scrawled in an effort to guess the answer to the problems of the day.

This reticence among the physics community to understand his work was on particular display at the first opportunity Feynman had to lecture about his idea, a few months later, at another National Academy of Sciences–sponsored workshop, the so-called Pocono Conference. His talk, titled "Alternative Formulation of Quantum Electrodynamics," came after Schwinger's, which lasted almost a full day. As unflappable as Schwinger was, even he got flustered from time to time, with Bohr, Dirac, and other big shots in the audience interrupting on a constant basis.

Hans Bethe noticed that Schwinger was interrupted the

least whenever his presentation became more formal, so he suggested Feynman make his presentation more formal and mathematical as well. This was like asking Bono to perform Bach on the harpsichord. Feynman had planned to present the material in almost the same way that he had worked on it, emphasizing successful calculations and results that would then take him backward to motivate the ideas. However, in deference to Bethe's suggestion, Feynman emphasized mathematics associated with his space-time sums rather than physics. The result was, in Feynman's words, "a hopeless presentation."

Dirac kept interrupting him to ask if his theory was *unitary*—a mathematical way of stating the fact, as I described earlier, that the calculated sum of the probabilities for all possible physical outcomes in any situation must equal unity (that is, there is a 100 percent probability that *something* happens)—but Feynman hadn't really thought about this, and since his particles were moving forward and backward in time, he answered that he simply didn't know.

Next, when Feynman introduced the idea of positrons as acting like electrons going backward in time, a participant asked whether this implied that in some of the space-time paths he was including in his calculations, several electrons might appear to be occupying the same state—a clear violation of the Pauli exclusion principle. Feynman answered affirmatively because in this case the different electrons were not really different particles, just the same particles going forward and backward in time. Feynman later recalled that chaos then ensued.

Ultimately this led Bohr to question the very physical basis of Feynman's space-time concept. He argued that the picture of space-time paths violated the tenets of quantum mechanics, which said that particles do not travel on individual specific trajectories at all. At this point Feynman gave up trying to convince his audience of the correctness of his scheme.

Bohr was wrong, of course. Feynman's sum over paths made it explicitly clear that many different trajectories must be considered simultaneously when calculating physical results, and in fact Bohr's son later came over and apologized, saying his father had misunderstood Feynman. But the type of questioning reflected what was clearly a deep skepticism and growing doubt that Feynman had developed a completely consistent picture. He was asking too much of his audience, which, while it contained some of the brightest physical minds in the twentieth century, couldn't have been expected to adjust in a single lecture to this totally new and still incomplete way of thinking about fundamental processes. After all, Feynman himself had taken years and thousands of pages of calculations to develop it.

One might think that Feynman would have been jealous of Schwinger, especially given the different early receptions to their work. There is no doubt that they were both competitive. But at least as Feynman recalled it, they were more like co-conspirators. Neither could fully understand what the other was doing, but both knew and trusted the other's abilities, and both felt they had left everyone else in the dust. Feynman's memory might have been self-serving, for he certainly was not content at the time with not being understood. Disconsolate after the Pocono Conference, he decided that he had to get his ideas into print so he could properly explain what he was doing. To Feynman, who hated writing up his work for publication, the motivation was clear.

Recall that Feynman's approach to solving physical problems could be framed as "The ends justify the means." By this I mean that one might strike out with a new half-baked idea or method, but its validity lay in the results. If the calculated results agreed with nature, via experimentation, then the method was probably on the right track and was worth exploring further.

Feynman sensed that his sum-over-paths approach was right. He had by then calculated almost every quantity one could calculate in QED and his results agreed with other methods when they were available. He built his methods as he went along, in order to address the specific questions at hand. But how could he present this in print to a physics audience used to not working backward, but forward to understand a theory?

To Feynman, getting an understanding of his ideas that was clear enough to present to others meant doing more calculations for himself. Over the summer of 1948 he both refined his computational methods and generalized them, and he developed even more powerful ones. By making the computational methods more succinct, and more general, he probably felt that he could communicate the results to the physics community more easily. Finally, by the spring of 1949 Feynman had struggled sufficiently that he had completed his epochal works, "The Theory of Positrons" and "Space-Time Approach to Quantum Electrodynamics," which essentially laid the basis of all of his ideas and his successful calculations performed over the previous two years.

Two other key factors contributed to cementing Feynman's resolve and his legacy. The first had to do with a remarkable young mathematician-turned-physicist we have already met, Freeman Dyson, who arrived in the United States in 1947 from the University of Cambridge as a student to work with Bethe, and who would eventually explain Feynman to the rest of the world.

Already famous in the United Kingdom for his mathematical accomplishments, at the age of twenty-three Freeman Dyson had decided that the truly interesting intellectual questions at that time lay in theoretical physics, particularly in the effort to understand the quantum theory of electromagnetism. So, while a fellow at Trinity College, Cambridge, he contacted several physicists and asked where he should go to catch up on the most exciting recent developments, and everyone pointed him to Bethe's group at Cornell.

Within a year, Dyson had completed a paper calculating the quantum corrections to the Lamb shift in a relativistic "toy" theory with spinless particles. Like Feynman, he was heavily influenced by his profound respect and admiration for Bethe, and his impression of the man was remarkably similar to Feynman's. As Dyson wrote, "His view was to understand anything meant to be able to calculate the number. That was for him the essence of doing physics."

By the spring of 1948 Dyson was focusing in depth on the conceptual problems of QED, and along with the rest of the physics world, alerted by Robert Oppenheimer, he read the first issue of the new Japanese journal *Progress in Theoretical Physics.* He was amazed to find that even though they were completely isolated, Japanese theorists had made remarkable progress during the war. In particular Sin-Itiro Tomonaga had essentially independently developed an approach to resolving the problems of QED using techniques similar to those developed by Schwinger. The difference was that Tomonaga's approach appeared to be far simpler to Dyson, who wrote, "Tomonaga expressed his in a simple, clear language so that anybody could understand it and Schwinger did not."

All during this time, Dyson was interacting with Feynman, learning at the blackboard exactly what he had accomplished. This gave him an almost unique opportunity to understand Feynman's approach at a time when Feynman had yet to publish his work, or even give a coherent seminar on the subject.

If there was anyone Dyson would grow to admire as much or more than Bethe, it was Feynman, whose brilliance, combined with energy, charisma, and fearlessness, was captivating to the young man. Dyson soon realized that not only was Feynman's space-time approach powerful, but if it were correct, it must be possible to unravel a relationship between this approach and the techniques Schwinger and Tomonaga developed.

At the same time, Dyson had so impressed his mentor that Bethe suggested he spend the second year of his Commonwealth Graduate Fellowship at the Institute for Advanced Study, with Oppenheimer. During the summer, before moving to New Jersey, Dyson went on that fateful cross-country car trip with Feynman to Los Alamos, then attended a summer school in Michigan, and took another cross-country trip, this time by Greyhound bus, to Berkeley and back. During the return trip from California, after forty-eight hours of what for many people would be mind-numbing bus travel, Dyson focused his thoughts intensively on physics, and was able complete in his mind the basic features of a proof that Feynman's approach and Schwinger's approach to QED were in fact equivalent. He was also able to fuse them together, as he described in a letter, in a "new form of the Schwinger theory which combines the advantages of both."

By October of 1948, before Feynman had completed his own epic paper on QED, Dyson submitted his famous paper "The Radiation Theories of Tomonaga, Schwinger, and Feynman," proving their equivalence. The psychological impact of this work was profound. Physicists had trust in Schwinger, but his methods were so complex as to be daunting. By demonstrating that Feynman's approach was equally trustworthy and consistent and provided a much easier and ultimately systematic method for calculating higher-order quantum corrections, Dyson exposed the rest of the physics community to an effective new tool that everyone could begin to use.

Dyson followed his "Radiation" paper with another seminal paper early in 1949. Having developed the methods to allow the adaptation of Schwinger's formalism to Feynman's methods in order to allow the calculation of arbitrarily complicated higher-order contributions to the theory, Dyson set himself the task of proving that it all made sense, in a rigorous fashion, or at least a fashion that was rigorous enough for physicists. He demonstrated that once the problems with infinities in the simplest self-energy and vacuum polarization calculations were resolved, then there were no other infinities that would result in the higher-order calculations. This completed a proof of what has now become known as the *renormalizability* of the theory, in which all infinities, once first controlled by mathematical tricks of the type that Feynman's method so easily allowed, can be incorporated in the unmeasurable *bare mass* and *charge* terms in the theory. When everything is expressed in terms of the renormalized physically measured masses and charges, all predictions become finite and sensible.

With the completion of Dyson's two papers, QED was truly tamed. It could now rise to the ranks of "theory" in the best scientific sense of being a logically consistent formalism that made unique predictions which could be and had been successfully compared with experimental results.

Interestingly, Dyson's original paper proving the equivalence of Schwinger's and Feynman's pictures contained only a single space-time diagram. Yet, as Feynman's seminal papers on his own results had not yet appeared, the first "Feynman diagram" in print was actually Dyson's.

It is hard to overstate the importance and impact of Dyson's work. He became almost an immediate celebrity in the physics community—but more importantly, his work led others, who could not understand or were dubious about Feynman's diagrammatic approach, to learn the formalism and begin to adopt it. While Feynman's own exposition of his work would come in seminal papers written in 1949 and 1950, it was Dyson's papers as much as anything that provided the window through which Feynman's ideas were able to change the way physicists thought about fundamental physics.

Dyson worked hard, perhaps harder than Feynman did, to convince the rest of the world of the utility of Feynman's ideas. Indeed, much of the subsequent use of Feynman diagrams by physicists in the decade following Feynman's work can be traced to the influence of Dyson, who, through his personal contact with Feynman, had become his number one disciple.

Dyson's papers ultimately reflected his view that while Feynman's and Schwinger's and Tomonaga's approaches were equivalent, Feynman's was the more enlightening and useful to those who would want to use QED to solve physics problems.

As the tide of interest began to shift away from Schwinger, he was not unaware, and later remarked, "There were visions at large, being proclaimed in a manner somewhat akin to that of the Apostles, who used Greek logic to bring the Hebrew god to the Gentiles." Schwinger later issued another backhanded compliment, which was perhaps even more characteristic of his view that enlightenment should be obtained only by those willing to toil the hardest: "Like the silicon chip of more recent years, the Feynman diagram was bringing computation to the masses."

The historian David Kaiser has studied the "dispersion" of the Feynman diagrammatic approach within the physics community in the years immediately prior to and following the publication of his work. As Kaiser has shown, the growth was exponential, with a doubling time of about two years, and by 1955 about 150 papers included Feynman diagrams. What began as a curiosity understood by Feynman and a few colleagues at Cornell, and then by way of Dyson and others at the Institute for Advanced Study and elsewhere, continued on to become a technique that has appeared since in every volume of the *Physical Review*, the standard reference journal in the field. What started out as an effort to handle QED is now used in virtually every field of physics.

Feynman and Dyson in fact had a very different understanding of his diagrams. Feynman, influenced by his own space-time pictures, his sum-over-paths formalism, and his insistence on thinking about particles in motion rather than quantum fields, thought of the diagrams as real pictures of physical processes, where electrons would bounce around from place to place and from time to time (both forward and backward). Based on these pictures he wrote down formulas, and he could then check to see if they would produce the correct answers. It really was a bootstrap approach, without much fundamental basis other than his remarkable intuition.

Dyson's work changed all that. He demonstrated how the diagrams could follow from fundamental sets of equations based on quantum field theory. For Dyson each part of each diagram represented a well-defined term in a series of equations. The diagrams were a crutch to help elucidate the equations, and the "Feynman rules" associated for translating the diagrams into equations were not ad hoc, as Feynman had invented them, but rather could be justified by well-defined manipulations of equations associated with quantum mechanics and special relativity. For this reason perhaps the adoption of Feynman diagrams occurred more quickly than the adoption of Feynman's path-integral space-time approach to physics—which, as we will see, would take another several decades to fully change the physics landscape.

Dyson had immediately appreciated how Feynman's

methods could help mere mortals systematically perform the complex calculations of quantum field theory. As he later described in a memoir, "The calculation I did for Hans Bethe using the orthodox theory took me several months of work and several hundred sheets of paper. Dick Feynman could get the same answer, calculating on a blackboard, in half an hour."

Feynman's own epiphany, when he knew he had discovered something truly outstanding, occurred during the January 1949 New York meeting of the American Physical Society, the very same meeting where Dyson was feted for his work by, among others, Robert Oppenheimer during his presidential address. At the time, when essentially all of the active research physicists in the country could fit in a single large hotel, the January event had become perhaps the most important physics meeting in the world. In 1949 the mood was particularly celebratory, as physicists had at last begun to see through the fog of infinities to understand electrodynamics as a workable quantum theory. Moreover, with the development of the first significant particle accelerator, Berkeley's 184-inch cyclotron, a plethora of new elementary particles interacting under the strange strong force were being created "under controlled conditions and in large numbers," as Feynman then described it in a review he wrote in 1948 for the new journal *Physics Today*. The excitement about this strange new world of phenomena, and the hope that the new methods that had been developed to deal with QED might shed light on it, were palpable. The catalyst for Feynman's enthusiasm came not from discussions about QED itself, but rather from a debate about the interactions of the newly created particles called *mesons*.

A young physicist, Murray Slotnick, had used the pre-Feynman methods of calculating in quantum field theory to determine, in a herculean effort, the effects these new types of particles might have if they were exchanged as virtual particles between neutrons in nuclei and electrons orbiting around them. He had discovered that only one type of possible interaction would produce finite results, which he then calculated. Slotnick presented his work at a session of the meeting.

After his presentation, Oppenheimer got up and, in his typical fashion, swiftly and harshly dismissed Slotnick's results. Oppenheimer claimed that a postdoctoral researcher at the institute had proved a general theorem that all of the different possible meson interactions would end up having the same effects on neutron-electron interactions, a result that was in manifest disagreement with Slotnick's claim.

Feynman apparently arrived after this session, but he was informed of the new controversy and asked to comment on his perception of who was correct. Up to this point Feynman had not done any theoretical calculations with mesons, but that evening, after someone explained to him what the theories were about, he translated his QED methods in this new context, and spent several hours calculating the various different possible meson theories. In the morning he compared his results with those of Slotnick, who had actually performed his calculation only in a special limit of the theory. Feynman had done the calculation in full generality, but when he took the same limit, his results agreed with Slotnick's.

Slotnick, who no doubt was reassured by Feynman's result, was also astonished. He had spent almost two years formulating and completing a calculation that Feynman did in a single evening. For Feynman, this realization was intoxicating. It was then, and only then, that he realized the true power of his new techniques. As he later said, "This is when I really knew I had something. I didn't really know that I had something so wonderful as when this happened. That was the moment that I really knew that I had to publish—that I had gotten ahead of the world. . . . That was the moment when I got my Nobel prize when Slotnick told me that he had been working two years. When I got the real prize it was really nothing, because I already knew I was a success. That was an exciting moment."

For good measure, when Oppenheimer's postdoctoral colleague gave his paper the next day, Feynman couldn't resist goading Oppenheimer. After the postdoctoral researcher in question, named Case, gave his presentation, Feynman got up and stated, offhandedly, that it had to be wrong because, as he now was able to say about Slotnick's work, "a simple calculation shows that it's correct."

This incident not only lit the fire necessary to convince Feynman he had to publish his results, but also forced him to develop the tools that allowed him to present his work in a fashion in which the rest of the community could understand it. In the first place, Feynman now felt obliged to understand what Case had done wrong. But to do that, he had to understand exactly what Case had done, which was difficult for him since Case had used the traditional techniques of quantum field theory that Feynman had ignored up to that point. In learning these techniques, from a graduate student at Cornell, Feynman was able not only to uncover Case's mistake, but also to reap an additional dividend from his investment of time: he was finally able to understand vacuum processes in a way that had earlier eluded him. His new understanding played a vital role in the presentation in the first of his epic QED papers, "The Theory of Positrons."

Next, with his newfound familiarity with meson theories, which were all the rage then, Feynman was able to formulate his diagrammatic rules for these theories and quickly reproduce every result that other physicists had derived over the years. These results were summarized in his next classic paper, "Space-Time Approach to Quantum Electrodynamics," no doubt raising its interest among the host of physicists who were struggling to make sense of these new strongly interacting particles.

As he had been urged at the Pocono Conference, where he had so deeply bungled his presentation, he decided to first publish his computational diagrammatic methods in a coherent fashion, and then the formal mathematical underpinnings later. Recognizing that his space-time approach "came from too far away," the introduction to his paper reads, "The Lagrangian method . . . was modified in accordance with the requirements of the Dirac equation and the phenomenon of pair creation. This was made easier by the reinterpretation of the theory of holes. Finally for practical calculations the expressions were developed in a power series. . . . It was apparent that each term in the series had a simple physical interpretation. Since the result was easier to understand than the derivation, it was thought best to publish the results first in this paper."

While minimizing the theoretical and mathematical motivations in this initial exposition, which he realized could not "carry the conviction of truth which would accompany the derivation," Feynman nevertheless made clear to the reader the significant advantage of his methods. When comparing his space-time approach with the more traditional "Hamiltonian" approach for dealing with the relativistic Dirac theory—as opposed to the nonrelativistic Schrödinger theory—he wrote, "As a further point, relativistic invariance will be self-evident. The Hamiltonian form of the equations develops the future from the instantaneous present. But for different observers in relative motion the instantaneous present is different, and corresponds to a different 3-dimensional cut of space-time. . . . By forsaking the Hamiltonian method, the wedding of relativity and quantum mechanics can be accomplished most naturally."

Over the next year, Feynman completed the remaining two papers necessary to give a thorough and more formal derivation of his results—using a new type of calculus he had developed—and to demonstrate formally the equivalence of his techniques with those of conventional quantum field theory. With the publication of Feynman's four papers, the QED story was now essentially complete. What had started out as a desire, when he was a graduate student, to reformulate QED as a theory without any infinities, had instead developed into a precise and astoundingly efficient methodology for working around infinities to obtain results that could be compared with experiments.

The contrast between hope and reality was not lost on Feynman. Throughout his four papers, while acknowledging the achievement they represented, he conveyed a palpable sense of disappointment. In "Space-Time," when comparing his computational techniques with Schwinger's, for example, he wrote, "Although in the limit the two methods agree, neither method appears to be thoroughly satisfactory theoretically. Nevertheless it does appear that we now have a complete and definite method for the calculation of physical processes to any order in quantum electrodynamics."

For many years, right up to and including the time, in 1965, when he was awarded the Nobel Prize for his work, Feynman felt that his methods were merely useful, not profound. He had not unveiled fundamental new properties of nature that would rid the theory of infinities, but had found a way to safely ignore them. The real hope—that path integrals would produce the revelations in our basic understanding of nature that he had hoped would cure the ills of relativistic quantum physics—he felt, had not materialized. As he said to a student newspaper on the day the Nobel Prize was announced, "It was the purpose of making these simplified methods of calculating more available that I published my paper in 1949, for I still didn't think I had solved any real problems. . . . I was still expecting that I would some day come through the other end of my original idea . . . and get finite answers, get that self-radiation out and the vacuum circles and that stuff straightened out . . . which I never did." And as he later described it in his Nobel address: "This completes the story of the development of the space-time view of quantum electrodynamics. I wonder if anything can be learned from it. I doubt it."

History would ultimately prove otherwise.

PART II

The Rest of the Universe

> Today we cannot see whether Schrödinger's equation contains frogs, musical composers, or morality—or whether it does not. We cannot say whether something beyond it like God is needed, or not. And so we can all hold strong opinions either way.
>
> —Richard Feynman

CHAPTER 11

Matter of the Heart and the Heart of Matter

> What I am trying to do is bring birth to clarity, which is really a half-assedly thought-out pictorial semi-vision thing.
>
> —Richard Feynman

Richard Feynman did not lay down his sword in 1949. He had achieved something wonderful, but even he didn't realize its full extent. In the meantime, nature continued to beckon. And for Feynman, being interested in physics meant being intensely curious about *all* aspects of the physical world, not just a single exciting frontier. The completion of his magnum opus on QED also coincided with a more personal and more difficult transition for him, from his beloved Cornell to the excitement and allure of more exotic climes.

As unconventional as Feynman was in his behavior and his way of dealing with social norms, he remained in those postwar years strangely provincial. I have already mentioned his early disinterest in classical music and the arts. And while he might have seemed worldly, largely through his wartime experience and his dealings with colleagues from other countries, by the age of thirty-one he had never ventured outside the shores of the continental United States. That would soon change.

Feynman's mind was always searching for new problems and new intellectual challenges. This predilection crept into his personal life as well. He once said to me, "One should seek out new adventures wherever one can."

I suspect that after the intense effort between 1946 and 1950 wrestling with QED, Feynman's mind now wanted to move far afield, not just in topic, but in style. Emotionally restless and unhappy, and yearning for adventure, Feynman wanted to escape the confines of gloomy wintery Ithaca, and perhaps the many tense sexual entanglements he had gotten embroiled in. Sunny California beckoned, but South America seemed even more enticing.

Like many key life decisions, this one involved a bit of serendipity. An old Los Alamos colleague, Robert Bacher, was moving to Pasadena, to rebuild a dormant physics program at Caltech, and he immediately thought of Feynman. He called him at the right time. The man who had previously turned down Princeton, University of Chicago, Berkeley, and a host of other institutions agreed to visit Pasadena. At the same time, Feynman's imagination was drifting even further. He had decided, for some reason, to visit South America, and had started to learn Spanish when a visiting Brazilian physicist invited him to Brazil for the summer of 1949. Feynman quickly accepted, got a passport, and switched to learning Portuguese.

He lectured on physics at the Centro Brasiliero de Pesquisas Fisicas in Rio de Janeiro and returned to Ithaca in the fall, more knowledgeable about Portuguese and the ways of Brazilians, thanks to a Copacabana beauty, or *garota*, named Clotilde, whom he had persuaded to accompany him back to the United States for a short while. The winter in Ithaca convinced him that he had to leave, and Caltech, in addition to better weather, had the appeal of not being a liberal arts university like Cornell, where, he said, "the theoretical broadening which comes from having many humanities subjects on campus is offset by the general dopiness of the people who study these things." He accepted the offer from Caltech and negotiated a deal that gave him the best of all possible worlds. He could take an immediate sabbatical year and head back to his beloved Brazil, where he could keep in touch with physics while swimming off the Copacabana beach and frolicking at night, all courtesy of Caltech and with support from the U.S. State Department.

His prime interest during this time was in the newly discovered mesons, and the confusion they introduced into nuclear physics. He used a ham radio as well as letters to contact his colleagues in the United States, and ask them questions, or issue advice. Fermi chided him: "I wish I could also refresh my ideas by swimming off Copacabana."

But Feynman also took seriously his own mission of helping to rejuvenate physics in Brazil. He taught courses at the Centro Brasileiro de Pesquisas Fisícas and chastised the Brazilian authorities for teaching students to memorize names and formulas but not to think about what they were doing. He complained they were learning how to explain words in terms of other words, but actually understood nothing and had no feel for the actual phenomena they were supposedly studying. For Feynman, understanding meant being able to take one's knowledge and apply it to new situations.

As brilliant as Feynman was, though, the isolation in Brazil kept him from keeping up with the forefront of the field at the time. He managed to independently reproduce results that had already been derived, but did not push the emerging field of particle physics forward. Instead, he had a cultural awakening and a sexual feast.

First, music. Feynman claimed he was tone-deaf, but there is no doubt, even if he marched to the beat of a different drummer, that he was born with rhythm. All those who were close to him knew that he was constantly drumming with his fingers whenever he worked, on paper, on walls, on anything that was convenient. In Rio, Feynman found the perfect music for his psyche—samba, a hot, rhythmic, and unpretentious hybrid of Latin and African traditions. He joined a samba school and began drumming in samba bands. He even got paid for his efforts. The peak occurred during the annual Carnaval, a debauched street festival, where he could carouse with abandon. And carouse he did. (Purely by coincidence I am writing this as I stare out from a hotel at Copacabana beach.)

It is easy to understand the fascination Rio had for Feynman. The city is breathtakingly beautiful, surrounded by gorgeous mountain and ocean scenery, and vibrant with the Rio *cariocas*, locals occupied with partying, arguing, playing soccer, and flirting on the beach. The full spectrum of human activity is almost always on display. The city is seedy, sexy, intense, scary, friendly, and relaxed, all at the same time. There Feynman could escape the confinement of a university town, where one could never quite get away from colleagues or students. Moreover, the Brazilians are a warm, friendly, and accepting people. Feynman could blend in. His own intense enthusiasm, never far from the surface, must have resonated with everyone around him—physicists, local *cariocas*, and, naturally, women.

Feynman lived at the Miramar Palace Hotel on Copacabana beach, where he descended into his lonely orgy of drinking (until he frightened himself enough to swear off alcohol for good) and sex. He picked up women on the beach and in clubs and at the hotel's patio bar, whose proximity to the action of Copacabana was, and still is, addictive. For a while he specialized in stewardesses who stayed at the hotel, and as he famously described over and over again, he enjoyed outsmarting the local women he met in bars. He convinced one of them not only to sleep with him, but also to repay him for the food he had bought her at the bar.

As often happens, however, this anonymous sex, while diverting, only reinforced his detached loneliness, and perhaps that is why he committed an utterly ridiculous and out-of-character act. He proposed, by letter, to a woman in Ithaca he had known and dated, a woman so different from the rest, and so different from Feynman, that perhaps he convinced himself she was the perfect complement.

Many of his previous girlfriends realized that the mutual enjoyment they thought they were sharing with Feynman was not being fully reciprocated. Feynman could concentrate completely on a woman he was with, in a way that was utterly captivating. But at the same time, as intense as his physical participation might have seemed, he was really alone with his thoughts. Mary Louise Bell, not savvy to this flaw, apparently pursued him from Ithaca to Pasadena. A platinum blonde with a penchant for high heels and tight clothes, she somehow thought that with Feynman she had the scaffolding from which she could fashion a final structure to her liking, one with a more polished exterior and a better appreciation of the arts, and one who wouldn't hang around with so many scientists.

They married in 1952. With hindsight some have said that divorce was inevitable, but there are no real rules from which one can make accurate predictions in matters of the heart. Nevertheless, one of the items brought up in the divorce proceedings was telling. She reported, "He begins working calculus problems in his head as soon as he awakens. He did calculus while driving his car, while sitting in the living room and while lying in bed at night."

During the first years together, as he settled into Pasadena following his wild year in Brazil, and his domestic bliss slowly turned into another private hell, he began to think he had made a mistake not only in choice of companion, but in choice of locale. He even wrote to Hans Bethe to discuss moving back to Cornell. But Caltech's lure was greater than Mary Louise's, and four years after their marriage, in 1956, he and Mary Louise parted ways, but he remained in Pasadena.

His new university was quickly growing to become a rival to his own eastern alma mater, MIT. It was an institution that, with its growing experimental and theoretical prominence in fields ranging from astrophysics to biochemistry and genetics, combined with the practical leanings of an engineering school, seemed like a perfect fit. It was. He would stay for the rest of his life.

Physics was experiencing a period of turmoil at the same time as Feynman's personal upheavals. Newly discovered elementary particles, mesons and the like, were proliferating madly in the newly built particle accelerators. The elementary particle physics zoo was becoming embarrassingly crowded, so crowded in fact that it wasn't clear which of the new blips on chart recorders and new tracks in bubble chambers might really represent new elementary particles and which were simply rearrangements of existing ones.

While Feynman had dabbled early on in the theory of mesons when he was perfecting his understanding of QED, he was also smart enough and realistic enough to know that his new diagrammatic methods were inappropriate to the task at hand. Not only were many of the experiments inconclusive, but the interactions between particles were generally so strong that the systematic effort to use Feynman diagrams to calculate small quantum corrections to processes seemed misplaced. He wrote to Enrico Fermi from Brazil: "Don't believe any calculation in meson theory that uses a Feynman diagram!" Elsewhere he referred to the field of meson physics by saying, "Perhaps there aren't enough clues for even a human mind to figure out what is the pattern."

I suspect that, in his view, the experimental world of mesons wasn't yet ready for interpretation, and he had a desire to strike out in a new intellectual direction, one that wasn't governed so much by attempting to unravel the mathematical intricacies of the quantum world as much as directly trying to puzzle out its physical consequences.

He wanted to think about something he could feel and play with, and not something he could only see in his mind. Thus, shortly after arriving at Caltech, Feynman turned to a completely different problem in a different area of physics. He began to explore not the quantum world of the very small, but the very cold.

The Dutch physicist Kamerlingh Onnes, who worked during the end of the nineteenth and early twentieth centuries, devoted his entire professional life to the physics of the very cold, cooling down systems closer and closer to absolute zero, the temperature where, classically at least, all internal motions of atoms would stop. In so doing, he made a miraculous discovery in 1911. At a temperature of 4 degrees above absolute zero (Onnes eventually got to less than 1 degree above absolute zero, reaching the coldest temperature ever achieved on earth up to that time), he witnessed a spectacular transition in mercury, in which electrical currents suddenly appeared to flow without any resistance at all.

It had been speculated that electrical resistance would decrease at very low temperatures, based on the simple observation that such a decrease was also observed at higher temperatures. Onnes himself speculated that the resistance would drop to zero at absolute zero, a temperature that can never be obtained directly in the laboratory. However, his amazing result was that the resistance abruptly dropped to exactly zero at a finite small, but nonzero, temperature. In such a state, an electric current, once started, would never stop. Onnes had discovered the phenomenon he called *superconductivity*.

Interestingly, when Onnes won the Nobel Prize two

years later, he did not win it explicitly for this discovery, but rather for his general "investigations on the properties of matter at low temperatures which led, inter alia, to the production of liquid helium." The prize showed unusual prescience (actually dumb luck) on the part of the Nobel Committee because it turned out, for reasons no one could have suspected in 1913, that liquid helium itself had properties at least as fascinating as those related to the conductivity of mercury and other metals at low temperatures. In 1938 it was discovered that liquid helium, when cooled sufficiently, exhibits a phenomenon known as *superfluidity*, which on its surface seems even stranger than superconductivity. Again, equally remarkably, Onnes probably cooled liquid helium to temperatures where it was superfluid, but didn't remark on this otherwise remarkable phenomenon.

In its superfluid phase, helium flows with no friction whatsoever. Put it in a container, and it will spontaneously flow in a thin film up over its edges. No matter how small a crack, it will flow through it. Unlike superconductivity, where the magic is hidden behind resistance and current measurements, with superfluidity it is on full display before our eyes.

As late as the early 1950s neither of these remarkable phenomena had yet been explained in terms of a microscopic atomic theory. As Feynman put it, they were like "two cities under siege . . . completely surrounded by knowledge although they themselves remained isolated and unassailable." At the same time, he was enamored with all of the fascinating new phenomena that nature revealed at low temperatures, and said, "I imagine experimental physicists must often look with envy at men like Kamerlingh Onnes, who discovered a field like low temperature, which seems to be bottomless and in which one can go down and down." Feynman was fascinated by all of these phenomena, but he turned his attention primarily to the mysteries of liquid helium, although he continued to struggle, ultimately unsuccessfully, to unravel the origin of superconductivity.

While at the time the field of what was eventually to become known as *condensed matter physics* was small, it is still hard to overemphasize the dramatic jump that Feynman had decided to take. Although the problems of superconductivity and superfluidity had not yet been solved, the people working in the field included some of the best minds in physics, and they had been thinking about the problems for some time.

Feynman clearly thought a fresh approach was needed, however, and among all of his research efforts, perhaps none better demonstrated how his remarkable physical intuition, combined with his mathematical prowess, could go around, rather than break down, preexisting barriers to understanding. The physical picture he ultimately derived achieved all of his goals for understanding superfluidity and, after the fact, seems remarkably simple—so simple that one wonders why no one else had thought of it. But that was a characteristic of Feynman's work. Beforehand everything was mired in mist, but after he had shown the way, everything seemed so clear as to be almost obvious.

Besides his general fascination with interesting phenomena in physics, one might wonder what it was about the problems of liquid helium in particular, and the applications of quantum mechanics to the properties of materials in general, that caused him to shift his focus to this area. I suspect that once again, the motivation might have come from his early efforts to understand the properties of the new strongly interacting elementary particles, mesons. Whereas he realized that Feynman diagrams were not likely to help unravel the thoroughly confusing experimental situation associated with the plethora of new strongly interacting particles emerging from accelerators, he nevertheless was interested in other physical ways that one might quantitatively understand the relevant physics that governed other strongly interacting systems.

The properties of electrons and atoms in dense materials provided for him precisely a similar problem, but one in which the experimental situation was much cleaner and the theoretical landscape at the time much less crowded. Indeed, until Feynman approached the problem, no one had attempted to use quantum mechanics at a microscopic level to directly derive the general properties of the transition of liquid helium from a normal state to a superfluid state.

That quantum mechanics played a key role in superfluidity was clear early on. In the first place, the only systems known in nature that behaved in a similar way, with no dissipation and no loss of energy, were atoms. According to the laws of classical electromagnetism, electrons orbiting in circles around protons should lose energy by radiation, so that the electrons would quickly spiral in to the nucleus. However, Niels Bohr postulated, and Erwin Schrödinger eventually demonstrated with his wave equation, that electrons could exist in stable energy levels where their properties would remain fixed in time, with no dissipation in energy.

So much for individual electrons or atoms, but could a whole macroscopic system like a visible amount of liquid helium exist in a single quantum state? Here there was another clue quantum mechanics was important. Classically, absolute zero is defined as the temperature where all motion ceases. No heat energy exists for atoms to vibrate or jostle one another, as in a standard gas or liquid, or even a solid. Moreover, assuming separate helium atoms had some small residual attraction with each other, which is required so a liquid simply doesn't fall apart into a gas of individual atoms, then at absolute zero, or near it, helium liquid should instead freeze into a rigid solid, with atoms held in place by their mutual attraction and no heat energy to move them around.

However, this isn't the case. As cold as anyone can make it, down well below 1 degree above absolute zero, as even Onnes showed, helium doesn't solidify. Quantum mechanics is once again the culprit. Even the lowest energy state in any quantum system always has a nonzero energy, associated with quantum fluctuations. Thus, even at absolute zero, helium atoms would still jiggle around. Helium is very light while at the same time the attraction between helium atoms is small enough so that the quantum ground-state energy of the atoms is sufficient enough to cause them to overcome this attraction and move about in a liquid form, rather than freezing in a lattice like a solid. Hydrogen atoms, which are even lighter, would exhibit the same phenomenon, except that hydrogen atoms are much more strongly attracted to each other, so their ground-state energy at low temperatures is not sufficient to break apart the bonds of a solid, and hydrogen freezes.

Helium is therefore unique in remaining a liquid at low temperatures, and its uniqueness is inherently quantum mechanical in origin. Thus, it makes sense that quantum mechanics also governs the transition that turns helium from a normal liquid to a superfluid at about 2 degrees above absolute zero.

As early as 1938, the physicist Fritz London had suggested that the transition to superfluidity might be a macroscopic example of a phenomenon that Einstein and the Indian physicist Satyendra Bose had predicted in an ideal gas of bosons—that is, particles with integer values of spin. Unlike fermions, which, as I have described, are subject to the Pauli exclusion principle, and cannot be in the same state at the same time and place, bosons behave precisely the opposite. As Bose and Einstein predicted, a gas of bosons can, at sufficiently low temperature, *condense* into a single macroscopic quantum state, where all of the particles are in precisely the same quantum state, and the macroscopic configuration would behave as a quantum mechanical and not a classical object. (In a classical object the individual particles have probability amplitudes that are completely uncorrelated with those of their neighbors. As a result, any fancy quantum interference between particles, which comes about by exact cancellations of different probability amplitudes, and which produces many of the strange aspects of the quantum world, is lost.)

The problem, however, is that *Bose-Einstein condensation*, as it is called, occurs for an ideal gas, in which the individual particles have no interactions with each other. Helium atoms, however, have weak attraction at a distance, and strong repulsion when they are very close. Was it possible that such a system could still have a Bose-Einstein–like condensation transition? This was one of the problems that drew Feynman's interest.

He not only was interested but also had developed the tool that allowed him an intuitive understanding of quantum mechanical effects. His method of recasting quantum mechanics as a sum over paths, with each path weighted by its action, provided, he believed, the perfect framework for picturing the microscopic phenomena that governed liquid helium at low temperatures.

As Feynman began to think about the sum over paths for each particle in the quantum liquid, two key factors guided him. First, since the helium atoms are bosons, the quantum mechanical amplitude describing their configuration is independent of which boson is where—it remains unchanged if the positions of any two helium atoms are exchanged. This means that paths dominating the path integral (that is, those with the smallest action), in which individual particles returned to their same position, had to be treated as identical to paths where the final positions of all of the particles resembled the initial configuration, while some of the particles had interchanged positions with one another. On the surface this may seem like an irrelevant mathematical subtlety, but it turns out to profoundly affect the physics.

The second factor focused on the action associated with the motion of any one helium atom in the background of all of its neighbors. Remember that the classical action associated with any trajectory involves summing up the differences between the kinetic and potential energies at all points along the path. Feynman reasoned that as any helium atom moved along at some velocity, it could reach any other point without getting close to another helium atom and experiencing a large repulsive potential (which would increase the action of the path) as long as the neighboring helium atoms simply rearranged themselves to make room for the helium atom as it moved from one place to another. If the atom was moving slowly, then the neighboring atoms would only have to move slowly to get out of the way. In the process of moving out of the way, these helium atoms would gain kinetic energy that would contribute to the action, but their kinetic energy would depend on the speed, and hence the kinetic energy, of the first helium atom.

The net effect of this, Feynman reasoned, was simply to change what we would normally think of as the mass of the helium atom in question. This is because when it moved, more than one helium atom would have to also move out of the way along with it to minimize the action. Everything else would remain the same.

As a result, Feynman demonstrated that the trajectories that contributed most to the sum over paths—that is, the trajectories that had minimum action—would be those in which each particle moved along acting like a free particle, but with a slightly increased mass. The otherwise strong repulsive atomic interaction at small distances could be completely accounted for by this effect and thus otherwise ignored. But if the particles were acting like free particles, then a Bose-Einstein ideal gas picture was a good one, and a Bose-Einstein transition was indeed possible.

That Feynman was able to demonstrate that strongly interacting particles could nevertheless behave, from the point of view of calculating their quantum mechanical behavior, as if they were free particles, clearly had significance for him beyond the case of liquid helium. As he stated in his first paper on the subject, "This principle may have uses in other branches of physics, for example in nuclear physics. Here there is the puzzling fact that single nucleons often act like independent particles in spite of strong interactions. The arguments we have made for helium may apply to this case also." Feynman was clearly deeply moved by this phenomenon. We will see again and again how his work over the next twenty years or so returned to precisely this situation, where objects that may be strongly interacting can at the same time behave as if they are not.

His interest went beyond nucleons (that is, protons and neutrons) alone. Once again, he was interested in trying to understand the new strongly interacting mesons, where Feynman diagrams appeared to be inadequate. If he could use his physical intuition, combined with the wealth of experimental information on liquid helium, to test new methods to understand strongly interacting systems, maybe he could then apply them to studying mesons. He as much as said this in the next paper he wrote, in 1954. This paper did not involve helium per se, but rather explored the motion of slow electrons in materials that become polarized in their presence, once again using space-time path-integral methods to untangle the physics. Again, in his words, "Aside from its intrinsic interest, the problem is a much simplified analog of those which occur in the conventional meson theory when perturbation theory is inadequate." And then in another paper almost a decade later: "It is interesting as a phenomenon in solids, but it has an extended interest since it is one of the simplest examples of the interaction of a particle and a field. It is in many ways analogous to the problem of a nucleon interacting with a meson field. . . . It is the strong-coupling aspect of the problem which has aroused so much interest."

Clearly, while Feynman reveled in the unity of physics—the applicability of understanding phenomena in one realm of the physical universe to understanding phenomena in a distant realm—his continued references over these years to mesons and nucleons suggest that he was always drawn back to the mysteries of the burgeoning subatomic world of elementary particles and their exotic new interactions. He would shortly return to that world, but he had not yet fully solved the problem he had set out for himself—namely, explaining superfluidity and perhaps with it the holy grail of superconductivity—and it was not in his nature to leave a problem he had started before he got the answers he wanted. In so doing, he would help transform the way we understand the quantum behavior of materials.

CHAPTER 12

Rearranging the Universe

Resistance is futile.

—The Borg, to Captain Picard, in *Star Trek: The Next Generation*

Feynman had explained how the transition that led to superfluid helium could be understood as a Bose-Einstein–like condensation where all the atoms condensed into a single macroscopically visible quantum state. But that did not solve the problem. The world does not look quantum mechanical to us because the mysterious quantum mechanical correlations at the atomic level that produce all the weird phenomena can be destroyed quickly by interactions with the environment. As a system gets bigger and bigger, the number and variety of these interactions, including now internal interactions between the many constituents, increase, and "quantum coherence" is quickly lost on microscopic timescales. Thus, simply condensing into a macroscopic quantum state is one thing, but why doesn't the smallest disturbance destroy this state? What *keeps* superfluid helium a superfluid?

Up until Feynman began to work on this topic, the answers given to this problem were "phenomenological." In other words, since experiments clearly demonstrated that superfluidity existed, one could extract the general behavior of the system from experimental results and therefore infer what the microscopic physical properties of the system would have to be to reproduce those results. This may sound like a complete physical explanation, but it is not. Deriving microscopic physical properties from experiments is different from explaining why nature produces these properties. This was the goal Feynman set for himself, and he largely achieved it.

Lev Landau had proposed the correct phenomenological model. Landau dominated physics in the Soviet Union with a force of personality and breadth of interest that paralleled Feynman's, and Feynman respected him tremendously. Indeed, when in 1955 the Soviet Academy of Sciences invited Feynman to attend a conference, meeting Landau was one of the reasons Feynman initially jumped at the chance. Unfortunately, however, cold war tensions caused the State Department to advise him not to go, and he acceded to the request.

Unlike Feynman, whose intellectual "center" seemed to always move back to particle physics, Landau's remained on the physics of materials, the very area that Feynman was now focusing on. Landau had argued that the persistence of superfluidity implied that there are no other accessible low-energy states near the coherent Bose-Einstein condensate state at low temperature that disturbances could kick the quantum fluid into. A normal liquid has resistance to flow (that is, viscosity) because individual atoms and molecules bounce around hitting other atoms and molecules in the fluid, other impurities, or the container walls. These internal excitations only change the state of motion of individual atoms, but they dissipate energy from the fluid to the container and slow the flow of the fluid. However, if there were no new accessible individual quantum mechanical states for individual particles to be kicked into, then these particles could not change their state of motion as a result of any collisions. Thus the superfluid will continue to move uniformly, just as an electron orbiting an atom continues to do so without dissipating any energy.

What Feynman hoped to demonstrate on the basis of first principles in quantum mechanics, using his path-integral picture, is that Landau's conjecture was correct. Here he utilized the crucial fact that I described earlier—namely, that helium atoms are bosons, which means that the quantum mechanical amplitude describing a state of *N* helium atoms will remain the same if any sets of atoms merely interchange their positions.

As I alluded to earlier, Feynman's argument was deceptively simple. First, he argued that because of the short-range repulsion of helium atoms, the lowest-energy ground state of the liquid will be that of roughly uniform density. He thought of each atom as being confined to a "cage" given by the positions of all of its neighbors, each of which exerts a repulsion if the atoms get too close. If the density of the liquid is higher in some place, that would mean that the cage surrounding one of the atoms would be smaller, confining the atom in question to a smaller space. But the Heisenberg uncertainty principle tells us that confining the atom to a smaller space raises its energy. Thus the energy of the system will be lowest when all of the atoms are as far apart from their neighbors as they can be, with nearly uniform average density.

One low-energy state that always exists involves "sound waves" of very long wavelength. Sound waves are "density waves," which means the density varies slowly across the liquid, and the atomic forces that resist compression act like little springs, causing a density excess to travel at a speed, which defines the speed of sound, across the liquid. As long as their wavelength is very large, so the variation in density is very gradual, these sound waves cost very little energy. They also do not change the properties of the liquid, nor, more importantly, do they affect the liquid's flow.

But again, why do no other low-energy states in the liquid exist?

Recall that quantum mechanics implies that all particles can be thought of as probability waves—where the amplitude of the wave is related to the probability of finding the particles at various places. But they are not called wave functions without good reason. A general characteristic of waves in quantum mechanics is that the energy associated with the wave is determined by its wavelength. Wave functions that wiggle a lot over small regions have higher energies than those that don't.

The reason for this is closely related, in fact, to the Heisenberg uncertainty principle. If a wave function varies from a high value to a low value over a very small distance, then one can localize the particle that the wave function describes within a very small range. But that means the momentum, and hence the energy uncertainty associated with the particle, is large.

The key, then, to finding a quantum state with low energy is to have a wave function without many closely spaced wiggles. Now, as I have described, simply having long wavelength sound waves does not destroy superfluidity, so we must consider other possibilities. Feynman said, let us start with the ground-state configuration, whatever that is, with uniform density, and imagine how we can create a state that differs from it, but only over large distances, so any wiggles in the wave function will not be closely spaced. We might imagine achieving this by moving some individual atom, *A*, a long distance away, to some new position *B*. But if the new configuration is also to be one of uniform density, the other atoms must rearrange themselves, and some other atom must move to take the place of the first atom.

Now, having moved the atom over a long distance, we might think this state differs significantly from the first one only over long distances because particle *A* has been displaced. But, as Feynman pointed out, all helium atoms are identical particles, and bosons to boot. Therefore, even though the displacements have been large, since the end result has simply been an interchange of identical bosons, this does not represent a new quantum configuration.

Think about this example for a while, and you will realize that no matter how far we move a particle, the new wave function for the system that will result can never reflect displacements of the particle that vary by more than about half the average distance between neighboring particles. Any motion by a greater distance can be replaced by a set of interchanges of other identical helium atoms, which won't change the wave function at all.

This means that the largest additional wiggles that we can introduce into the wave function to describe a new state of the system cannot be bigger than the average inter-atomic spacing. But wiggles of this scale or smaller correspond to excited energies that are relatively high, certainly far higher than random thermal fluctuations could produce at the low temperatures where superfluidity is observed.

Thus, Feynman demonstrated by this elegant physical reasoning that the statistics of bosons directly implied that there are no low-lying excited states above the ground state that could be easily accessible by the motion of atoms so as to produce a resistance to current flow. The superfluid ground state would persist as long as the thermal energy available to the system was smaller than the gap between the ground state and the lowest-energy excited state.

He did more, of course. Using his path-integral formalism he was able to estimate the energy of the excited states, which Landau had called *rotons*, by varying all reasonable guesses for the wave functions and calculating the state of minimum energy. The approximations were crude and initially did not match the existing data that well, but over the course of the decade he refined his analysis and made predictions that agreed well with the data.

Before Feynman had started to work on liquid helium, the physicist László Tizsa, whom I later knew only as a delightful retired professor at MIT, had proposed what he called a *two fluid model* to describe how the transition between superfluid liquid helium and the normal fluid would take place. Landau later extended the idea. He imagined that at absolute zero temperature all of liquid helium would be in the superfluid state. Then as the fluid was heated, some excitations would be created and would move about in the background superfluid, but could collide with walls and dissipate energy, acting therefore like a normal fluid component. As the system is heated more, more excitations are created until finally the normal fluid component fills the full volume.

Feynman's quantitative first-principles estimates again reproduced the general physical picture, but it would take thirty-two years before sufficiently detailed calculations could be done to obtain good agreement with data. In 1985, using a supercomputer to carry out the detailed path integrals that Feynman had roughly approximated, physicists were able to verify that this method could produce excellent agreement with the detailed nature of the liquid helium transition between the normal and superfluid phases.

But perhaps the most impressive bit of physical prestidigitation that Feynman pulled out of his sleeve related to solving the following problem: what would happen to a bucket of superfluid helium if one twirled it? As with many physics problems, this might not seem like a pressing issue, until you think about it. Feynman pointed out that because of the nature of the ground state and the energy required for excitations above it, the superfluid state had to be "irrotational," which meant that no eddies could form that might impede current flow. But now what would happen if we caused the entire fluid to rotate by rotating the container it was in? Feynman worked out the key to what would happen, and he had no idea at the time that the Nobel Prize–winning Norwegian-American chemist Lars Onsager had also suggested a similar solution. Once again, the rules of quantum mechanics played a key role.

Recall that Niels Bohr himself had started the quantum revolution by hypothesizing that as an electron orbited an atom, only certain energy levels were allowed. The energies were thus quantized. But the principle that determined the quantization rule actually derived from the orbital angular momentum of the electron around the atom. Bohr hypothesized that this angular momentum was quantized in terms of multiples of some smallest unit (the same unit I described earlier for electrons, in which their spin angular momentum is ½).

If the superfluid is also to be governed by quantum mechanics, its orbital angular momentum, were it to be forced to circulate around, would also be quantized in terms of the same fundamental unit. This would mean that there would be a minimum amount of circulation.

Feynman thought long and hard about how to minimize the energy in such a fluid, and he ultimately came upon a physical picture in which the fluid as a whole would not rotate, but many small regions—as small as possible in fact, on the order of several atoms across—would rotate, each around its own central region. These central regions would line up in the vertical direction to form vortex lines, like the funnels of a tornado, or the swirling water around an emptying drain (as Feynman put it). These vortex lines would distribute themselves throughout the background, nonrotating fluid, with uniform density.

Thinking about vortices allowed Feynman to estimate many facets of the behavior of liquid helium, including how resistance would set in as vortex excitations are produced. These vortices would twist and tangle around each other as the fluid flows, and at a velocity that was a hundred times smaller than one would otherwise imagine, superfluidity would be destroyed.

Feynman's creative imagining ultimately produced a very physical picture of Landau's rotons, those lowest-energy local excitations. Feynman realized that a vortex need not line up from the top of a container to the bottom, but it could also curl up upon itself in a ring form. He thought of the smoke rings he had studied as a high school student, and realized that perhaps the quantum version of the smallest such ring of atoms that could exist in the liquid might describe a roton, from which he derived its properties. As he worked through the mathematics in his head and tried to solve the equations to figure out how to match the intuitive picture of a smoke ring with the mathematics of the Schrödinger equation, he ultimately came up with another very physical picture: that of a continuous ring of schoolchildren quickly going one by one down a slide, then slowly heading back to climb up the ladder before going down again. The roton could be a local region where the fluid was moving at a different speed relative to the background fluid, but in order for the angular momentum properties of quantum mechanics to hold, the fluid would have to flow backward again somewhere else, like a vortex. And by curling up in a ring, the vortex would shrink until it carried the smallest possible energy, the energy of a roton.

All of these musings are amusing, but what is particularly important about them is how they once again changed the way physicists in this field thought about their subject. Feynman's intuitive guesses for "test wave functions" that he could vary to explore which ones had minimum energy established the use of what is called a *variational method* in condensed matter physics, and one that has subsequently been used to address almost all of the key outstanding problems in the study of matter in the last half century.

Feynman missed out on solving some of them, but the

influence of his ideas was unmistakable. Take the "one that got away," superconductivity. Feynman never succeeded in obtaining the physical breakthrough that the physicists John Bardeen, Leon Cooper, and Robert Schrieffer did to explain this phenomenon—largely because Feynman did not attempt to fully follow previous work in the field, a nagging characteristic that would cause him to miss out on a number of key discoveries—but their approach borrowed heavily from the ideas he introduced to study the properties of materials in general, and specifically the ideas he developed to explain superfluidity. Feynman's first paper applying his space-time approach to understanding the properties of electrons in materials detailed the importance of the electron coupling to vibrational modes in the material. This coupling turned out to be of crucial importance in understanding the interactions that allow electron pairs to bind together and condense in a superconductor. Indeed, a year before he and his colleagues finally cracked the problem that led to their own Nobel Prize, Schrieffer was in the audience listening to Feynman talk about both superfluidity and superconductivity, and was fascinated to hear Feynman talk in great detail about his own ideas about superconductivity that went wrong. Their own approach to understanding superconductivity was to figure out, just as Feynman had, how a Bose-Einstein–like condensate could form, in this case for particles like electrons that were not bosons. Equally important was demonstrating, as Feynman had demonstrated, that there was an energy gap between the ground state and the excited states, so that at low energies, collisions which otherwise produce excitations that dissipate energy could not occur.

Feynman's ideas about how angular momentum would be introduced into a superfluid that was forced to rotate, through the formation of vortices, was also remarkably prescient. A precisely similar phenomenon occurs in superconductors. Here, a superconductor will normally not allow a magnetic field to exist inside of it, just as a superfluid will tend not to move in circular eddies. However, as Feynman showed, if the fluid is forced to rotate (say, by putting it under high pressure and freezing it, and then rotating it and allowing it to melt), the circulation will occur by the production of vortices. Alexei Abrikosov later showed that one can force magnetic field lines through a superconductor, but they too will permeate the superconductor in thin vortex lines. He won the Nobel Prize for his work and said, in his Nobel address, that he put his original proposal in a drawer because Landau didn't think much of it. It was only after learning about Feynman's thoughts about rotational vortices in superfluids that Abrikosov had the courage to publish his ideas on magnetic vortices. Feynman's foray into condensed matter physics was thus remarkable, not just for the manner in which his intuition led to key insights, but by the way in which, in the course of less than a half-dozen pieces of work, his imprint on the field was demonstrable.

During this period, Feynman enjoyed reaching out to a new community of physicists, but he also suffered the anxieties that occur when entering a field outside one's expertise. Sometimes you get stepped on. For example, after Feynman surprised Onsager's students with his predictions of vortices at a conference to which he was invited to speak, he had occasion to meet Onsager at a condensed matter physics meeting. At dinner before Feynman gave a talk, Onsager asked him, "So you think you have a theory of liquid helium?" Feynman answered, "Yes, I do." Onsager merely replied, "Hmpf," leading Feynman to think that Onsager didn't expect much from him.

The next day, however, when Feynman said there was one aspect of the phase transition he didn't understand, Onsager spoke up right away, saying, "Mr. Feynman is new in our field, and there is evidently something he doesn't know about it, and we ought to educate him." Feynman was petrified, but Onsager continued by saying that the thing that Feynman didn't understand about helium was not understood for any material, and "therefore the fact that he cannot do it for He II is no reflection at all on the value of his contribution to understanding the rest of the phenomena."

Feynman was so taken by this show of warmth that he and Onsager continued to meet and discuss things, even though Feynman did most of the talking. Onsager was not one of Garrison Keillor's Norwegian bachelor farmers from Minnesota, but he came from the same Norwegian stock, and spoke only rarely, when he felt there was something he really had to say.

And while Feynman may have missed out on a few extra prizes, the biggest one for him was always in understanding the physics. If he ever felt jealousy or regret because of a lost prize, or misplaced credit, he didn't show it. Numerous examples abound where he had worked something out to his own satisfaction, but had not felt the need to write up a paper, only to have someone else gain attention for the same idea. Likewise, if he discovered that someone else had a similar idea, he often referenced their own work rather than his own. For example, perhaps in response to Onsager's show of intellectual generosity at their first meeting, he always credited Onsager with the idea of vortices, even though he discovered Onsager's work long after he had derived his own results.

One of the best examples of Feynman's magnanimity in this regard occurred twenty years after he first started thinking about liquid helium and vortices. When thinking about two-dimensional systems like thin films of liquid helium Feynman realized that the possible appearance of vortices could dramatically change their properties, resulting in a type of phase transition that apparently violated a famous mathematical theorem about the behavior of such two-dimensional systems. He even wrote a paper on the subject. However, he discovered that two young, and then unknown physicists, John Kosterlitz and David Thouless, had just produced a similar paper on the same idea. Rather than swamp them, or scoop them, Feynman decided not to submit his paper for publication, being content on simply having solved the problem himself. The remarkable phase transition has become known as the Kosterlitz-Thouless transition.

But, much as Feynman enjoyed this respite from the unsettled experimental and theoretical terrain of particle physics, he still yearned to discover a new law of nature, and fundamental physics at the cutting edge was the best possibility. Thus, even as he was working on liquid helium, he was trying to keep up with what was going on in his home territory. Dirac had had his equation. Feynman didn't yet have his.

CHAPTER 13

Hiding in the Mirror

That was a moment when I knew
how nature worked.

—RICHARD FEYNMAN

The same mesons that had begun to invade Richard Feynman's world in 1950 had turned every particle physicists' world upside down by 1956. New particles kept being discovered, and each one was stranger than the one before. Particles were produced strongly in cosmic rays, but the same physics that produced them should have caused them to quickly decay. Instead, they lasted as long as a hundred-millionth of a second, which doesn't sound too long, but is millions of times longer than one would predict on the basis of first principles.

By 1956, Feynman's fame among physicists was secure. Feynman diagrams had become a part of the standard toolbox of the physics community, and anyone who was passing through Caltech would make every effort to pay homage. Everyone wanted to talk to Feynman because they wanted to talk about their own physics problems. The very same characteristic that made him so irresistible to women worked wonders on scientists as well. It was a characteristic he shared with the remarkable Italian physicist Enrico Fermi, the Nobel laureate who helped lead the Manhattan Project to build the first controlled nuclear reactor at the University of Chicago, and the last nuclear or particle physicist who was equally adept at theory as at experimentation. Fermi had developed a simple theory that described the nuclear process associated with the decay of the neutron into a proton (and an electron and ultimately a new particle that Fermi had dubbed a *neutrino*—Italian for "little neutron"), so-called *beta decay*, one of the essential processes that formed a part of the reactions that led to both the atomic bomb and, later, thermonuclear weapons.

Since the neutron lived almost ten minutes before decaying, a virtual eternity compared to the lifetimes of the unstable strongly interacting mesons that were being discovered in the 1950s, it was recognized that the forces at work governing its decay must be very weak. The interaction that Fermi developed a model for in beta decay was therefore called the *weak interaction.* By the mid-1950s it had become clear that the weak interaction was a wholly distinct force in nature, separate from the strong interactions that were producing all of the new particles being observed in accelerators, and that the weak force was likely to be responsible for the decays of all of the particles with anomalously long lifetimes. But while Fermi's model for beta decay was simple, there was no underlying theory that connected all of the interactions that were being observed and ascribed to this new force.

Fermi had built the University of Chicago theory group into an international powerhouse. Everyone wanted to be there, to share not just in the excitement of the physics but also in the excitement of working with Fermi. He had a highly unusual characteristic—one he and Feynman shared: When they listened, they actually *listened*! They focused all of their attention on what was being talked about, and tried to understand the ideas being expressed, and if possible, tried to help improve upon them.

Unfortunately, Fermi died in 1954 of cancer, probably induced by careless handling of radioactive materials at a time when the dangers involved were not understood. His death was a blow both to physics and to Chicago, where he had helped train young theorists and experimentalists who would come to dominate the entire field for the next generation.

After Fermi's death, young theorists began to be drawn to Feynman's magnetism. Unlike Fermi, he did not have the character or patience to diligently help train scientists. Yet, there was nothing so flattering as having Feynman direct all of his attention at their ideas. For once Feynman latched on to a problem, he would not rest until he had either solved it or decided it was unsolvable. Many young physicists would mistake his interest in their problems for an interest in them. The result was seductive in the extreme.

One man who was attracted from Chicago to Feynman's light was a twenty-five-year-old wunderkind named Murray Gell-Mann. If Feynman dominated particle physics in the immediate postwar era, Gell-Mann would do so in the decade that followed. As Feynman would later describe, in one of his characteristic displays of praise for the work of others, "Our knowledge of fundamental physics contains not one fruitful idea that does not carry the name of Murray Gell-Mann."

There is little hyperbole here. Gell-Mann's talents provided a perfect match to the problems of the time, and he left an indelible mark on the field, not just through exotic language like *strangeness* and *quarks*, but through his ideas, which, like Feynman's, continue to color discussions at the forefront of physics today.

Gell-Mann was nevertheless in many ways the opposite of Feynman. Like Julian Schwinger, he was a child prodigy. Graduating from high school at the age of fifteen, he received his best offer from Yale, which disappointed him, but he went anyway. At nineteen he graduated and moved to MIT, where he received his PhD at the age of twenty-one. He once told me that he could have graduated in a single year if he hadn't wasted time trying to perfect the subject of his thesis.

Gell-Mann not only had mastered physics by the age of twenty-one, he had mastered *everything*. Most notable was his fascination with languages, their etymology, pronunciation, and relationships. It is hard to find anyone who knows Murray who hasn't listened to him correct the pronunciation of their own name.

But Gell-Mann bore a distinct difference from Schwinger that led to his attraction to Feynman. He had no patience with those who couched their often-marginal efforts in fancy formalism. Gell-Mann could see right through the ruse, and there were few physicists who could be so dismissive of the work of others as he could be. But Gell-Mann knew from Feynman's work, and from listening to him talk, that when he was doing physics, there was no bullshit, no pretense, nothing but the physics. Moreover, Feynman's solutions actually mattered. As Gell-Mann put it, "What I always liked about Richard's style was the lack of pomposity in his presentation. I was tired of theorists who dressed up their work in fancy mathematical language or invented pretentious frameworks for their sometimes rather modest contributions. Richard's ideas, often powerful, ingenious, and original, were presented in a straightforward manner that I found refreshing."

There was another facet of Feynman's character, the showman, that appealed to Gell-Mann much less. As he later described it, he "spent a great deal of time and energy generating anecdotes about himself." But while this feature would later grate on Gell-Mann's nerves, following Fermi's death in 1954, as he considered where he might want to work if he left Chicago, and with whom he might want to work, the choice seemed clear.

Gell-Mann, along with the theoretical physicist Francis Low, had done a remarkable early calculation in 1954 using Feynman's QED that sought to address the question of what precisely would happen to the theory as one explored smaller and smaller scales—exactly the traditionally unobservable regime where Feynman's prescription for removing infinities involved artificially altering the theory. The result was surprising, and while technical and difficult to follow at the time, it eventually laid the basis for many of the developments in particle physics in the 1970s.

They found that due to the effects of virtual particle-antiparticle pairs that had to be incorporated when considering quantum mechanical effects in QED, physically measurable quantities like the electric charge on an electron depend on the scale at which they are measured. In the case of QED, in fact, the effective charge on the electron, and hence the strength of the electromagnetic interaction, would appear to increase as one penetrated the cloud of virtual electron-positron pairs that would be surrounding each particle.

Feynman, who was famous for ignoring the papers of others as he independently tried to derive, or more often re-derive, all physics results, nevertheless was very impressed with the paper by Gell-Mann and Low, and he told Gell-Mann so when he first visited Caltech. In fact, Feynman said it was the only QED calculation he knew of that he had not independently derived on his own. In retrospect, this was somewhat surprising because the effect of the Gell-Mann–Low approach was to lead to a totally different interpretation of removing infinities in quantum field theory than Feynman continued to express for some time. It suggests, as you will see, that Feynman's approach to renormalization, which he had always thought was just an artificial kluge that one day would be replaced by a true fundamental understanding of QED, instead reflected an underlying physical reality that is central to the way nature works on its smallest scales.

When Gell-Mann arrived at Caltech, it was clear to him, and to much of the rest of the physics community, that perhaps the two greatest physics minds in a generation were now located in a single institution. Everyone was prepared for spontaneous combustion.

While it is inevitably unfair to choose a single quality to characterize the work of such a deep and inventive mind, Gell-Mann had already begun to make his mark on physics, and would continue to do so, by uncovering new symmetries of nature on its smallest scales. Symmetry is central to our current understanding of nature, but its significance is vastly misunderstood in the public consciousness, in part because things that have more symmetry in physics are perceived to be less interesting in an artistic sense. Traditionally the more ornate the symmetries of a piece of artwork, the more it is appreciated. Thus, a beautiful chandelier with many identical curved pieces is treasured. The artwork of M. C. Escher, in which many copies of a fish or other animal are embedded within a drawing, is another example. But in physics, the symmetries that are most treasured are those that make nature the least ornate. A boring sphere is far more symmetric than a tetrahedron, for example.

This is because symmetries in physics tell us that objects or systems do not change when we change our perspective. Thus, for example, when we rotate a tetrahedron by 60 degrees along any of its sides, it looks identical. A sphere, however, has much more symmetry because when we rotate a sphere by *any* angle, no matter how small or large, it looks the same. Recognizing the fact that symmetries imply that something doesn't change as we change our perspective makes the significance of symmetries in physics, in retrospect, seem almost obvious. But it was not until the young German mathematician Emmy Noether unveiled what is now called *Noether's theorem* in 1918 that the ultimate mathematical implications of symmetries for physics became manifest.

What Noether—who was originally unable to get a university position because she was a woman—demonstrated was that for each symmetry that exists in nature, there must also exist a quantity that is *conserved*—namely, one that does not change with time. The most famous examples are the conservation of energy and the conservation of momentum. Often in school students learn that these quantities are conserved, but they are taught this as if it is an act of faith. However, Noether's theorem tells us that the conservation of energy arises because the laws of physics do not change with time—they were the same yesterday as they are today—and that the conservation of momentum arises because the laws of physics do not vary from point to point—they will be the same if we do experiments in London or New York.

The attempt to use the symmetries of nature to constrain or govern the fundamental laws of physics became more common in the early 1950s as the plethora of new elementary particles appearing in accelerators forced physicists to seek out some order amid the apparent chaos. The effort focused on the search for quantities that did not appear to change when one particle decayed into other particles, for example. The hope was that such conserved quantities would allow one to work backward to find out the underlying symmetries of nature, and that these would then govern the mathematical form of the equations that would describe the relevant physics. That hope has been realized.

Gell-Mann first became well known following his proposal in 1952 that the new mesons, whose production was so strong but whose decay was so weak, behaved this way because some quantity associated with the new particles was conserved in the strong interaction. He called this strange new quantity, perhaps not inappropriately, *strangeness.* The conservative editors of the *Physical Review*, however, where he first published his ideas, felt this new name was inappropriate in a physics publication, and refused to use it in the title of his paper on the subject.

Gell-Mann argued as follows: Because strangeness is conserved, the new particles have to be produced in pairs, particles and antiparticles, with equal and opposite values of this new *quantum number*. The particles themselves would then be absolutely stable, because to decay into non-strange particles would violate this conservation law—changing the *strangeness number* by one—if the strong force was the only one operating. However, if the weak force, the one responsible for the decay of neutrons and the reactions that power the sun, did not respect this conservation law, then the weak force could induce decays of these new particles. But because the force was weak, the particles would then survive a long time before decaying.

As attractive as this idea was, success in physics does not involve merely postdiction. What would it predict that could allow it to be tested? Indeed, this was the immediate reaction among many of Gell-Mann's colleagues. As Richard Garwin, a brilliant experimentalist who had played a key role in the development of the hydrogen bomb, put it, "I don't see what use it could possibly be."

The leap came when Gell-Mann realized that this strangeness quantum number could be used to classify sets of existing particles, and he even made a stranger prediction. He predicted that a neutral particle called the K-zero should have an antiparticle, the anti-K-zero, which was different from itself. Since most other neutral particles, like the photon, are equivalent to their antiparticles, this proposal was unusual, to say the least. But ultimately it proved to be correct, and the K-zero–anti K-zero system has served as a remarkable laboratory for probing new physics, cementing Gell-Mann's reputation among the then-rising new generation of particle physicists.

It was after his introduction of strangeness, and the death of Fermi, that Gell-Mann began to receive offers to work outside of Chicago. He wanted to go to Caltech to work with Feynman, and Caltech encouraged the decision by matching competing offers and making Gell-Mann, at age twenty-six, the youngest full professor in its history. The hope was that this would not be the first history-breaking event that Gell-Mann, along with Feynman, might bring to the university.

Feynman and Gell-Mann had a remarkable partnership involving intellectual give-and-take. The two of them would argue nonstop in their offices, a kind of friendly argument or, as Gell-Mann would later call it, "twisting the tail of the cosmos," as they tried to unravel the newest mysteries at the forefront of particle physics. It had an impact on their students and postdocs as well. I remember when I was a young researcher at Harvard, working with Sheldon Glashow, a former student of Schwinger's and a Nobel laureate. Our meetings were punctuated with a mixture of arguments and laughter. Glashow had been a postdoc at Caltech with Gell-Mann, and I expect was strongly influenced by the style of discussion he witnessed there, of which I, and hopefully my students, have become further beneficiaries. The partnership between Feynman and Gell-Mann also was an uneasy marriage of opposites. Gell-Mann was the very epitome of the cultured scientist, and Feynman was not. Gell-Mann was, by nature, judgmental of people and their ideas, and always worried about intellectual priority. Feynman had no patience for physics nonsense or pomposity and appreciated talent, but if he got scooped, as mentioned earlier, what he cared most about was whether he had been right or wrong, not who ultimately got credit. It was an interesting partnership, which, given the difference in character and style, was bound to run into trouble eventually—but not right away.

Nevertheless, this was a time when both scientists were near their creative peaks. Gell-Mann was just beginning to revolutionize the world of elementary particles, and Feynman had just completed his own revolution in quantum mechanics. When they began to work together, another vexing physics problem had arisen, also related in part to the new strange particles that Gell-Mann had been classifying. This problem was far more puzzling than the merely extra-long lifetimes that Gell-Mann's theory explained. It had to do with one of the most common, and commonsense, symmetries of nature that characterize the physical world.

At some point in our childhood we all learn to tell the difference between right and left. It's not easy, and Feynman used to tell his students that sometimes even he had to look at the mole on his left hand in order to be sure. That is because the distinction between left and right is arbitrary. If we called everything that we call left, right, and right, left, then what would change except the names? The real question is whether "left" and "right" are indeed merely human semantic constructs, or whether nature has a more fundamental way to distinguish them.

Think of it another way, along the lines that Feynman once described. If we were in communication with aliens on another planet, how would we tell them the difference between right and left? Well, if their planet had a magnetic field like earth's and orbited their star in the same direction as the earth does, we could have them take a bar magnet and align its north pole to point north, and then left could be defined as the direction in which the sun sets. But they would say, "Yes, we have a bar magnet, but which end is north?"

We could go on and on like this and convince ourselves that terms like *left* and *north* are arbitrary conventions we have invented, but that they have no ultimate meaning in nature. Or do they? Noether's theorem tells us that if nature doesn't change if we reverse right and left, there should be a quantity, which we call *parity*, that is conserved, that doesn't change no matter what physical processes are taking place.

This doesn't mean all individual objects are left-right symmetric, however. Look at yourself in the mirror. Your hair may be parted one way, or your left leg may be slightly longer than your right one. Your mirror doppelganger, however, has its hair parted in the other direction, and its right leg is longer. The things that remain identical, like a sphere, for example, when we flip right and left are said to have *even* parity, and those that change are said to have *odd* parity. What Noether's theorem tells us is that both even-parity and odd-parity objects would nevertheless obey the same laws of physics in a mirror world. The associated conservation law tells us that even-parity objects do not spontaneously turn into odd-parity objects. If they did, we could use this spontaneous transformation to define an absolute left or right.

Elementary particles can be classified by their parity properties, usually associated with the way they interact with other particles. Some have even-parity interactions and some have odd-parity interactions. Noether's theorem tells us that a single even-parity particle cannot decay into a single odd-parity particle plus an even-parity particle. It can, however, decay into two odd-parity particles because if one particle heads out to the left and one to the right, if the identity of the particles is also interchanged under such a parity switch at the same time as the directions of the particles are flipped as a result of interchanging left and right, then the outgoing configuration would look identical afterward—that is, it would have even parity, as the original particle had.

So far so good. However, physicists discovered that the decay of strange mesons called K-mesons—whose long lifetimes Murray had explained via strangeness—nevertheless was not obeying the rules. Kaons, as they are also called, were observed to decay into lighter particles called pions, but sometimes they would decay into two pions and sometimes into three pions. Since pions have odd parity, a state of two pions has different reflection properties than a state of three pions. But then it would be impossible for a single-type particle to decay into the two different configurations, because that would mean sometimes the initial particle would have even parity and sometimes odd parity.

The simple solution was that there must be two different types of kaons, with one type having even parity and decaying into two pions and one type having odd parity and decaying into three pions. The problem was that these two types of kaons, which physicists had dubbed the tau and the theta, otherwise looked completely identical. They had the same mass and the same lifetime. Why should nature produce two such identical but different particles? Various exotic new symmetries might be invented that would give them the same mass, and Gell-Mann and others had been pondering such possibilities, but to also produce the same lifetime seemed impossible, because the generic quantum probability to decay into three particles is much less than that to decay into two particles, all other things being equal.

This was the situation in the spring of 1956 when Feynman and Gell-Mann began to work together at Caltech, and both attended the major particle physics conference of its time, called the Rochester Conference, which was then still being held in Rochester, New York. There they heard about compelling new data that once again made the tau and theta look like identical twins.

The situation became so difficult to justify that some physicists privately began to wonder whether the tau and theta were distinct. At the conference, Feynman was rooming with a young experimental physicist, Martin Block. Records indicate that in the Saturday session near the end of the meeting, Feynman got up and raised a question for the experts that he attributed to Block, that perhaps the two particles might actually be the same, and that the weak interactions might not respect parity—that nature might, at some level, distinguish right from left.

Murray Gell-Mann was later described as having teased Feynman mercilessly afterward for not having had the courage to ask the question in his own name, so I contacted my old friend Marty Block and asked what really transpired. He confirmed that he had asked Feynman why parity couldn't be violated by the weak interaction. Feynman had been tempted to call him an idiot until he realized that he could not come up with an answer, and he and Marty debated the issue each night during the conference until Feynman suggested that Marty bring up this possibility at the meeting. Marty said no one would listen to him, and asked Feynman to raise it for him. Feynman ran the idea past Gell-Mann to see if he knew of any obvious reasons why this could not be possible, and he didn't. So Feynman, in his traditional way, was giving credit where credit was due, rather than avoiding ridicule by raising an outrageous and potentially obviously wrong possibility.

Feynman's question received a response from a young theorist, Chen Ning (Frank) Yang, who, according to the official report, answered that he and his colleague Tsung-Dao Lee had been looking into this issue but had not reached any conclusions. (Block told me that the report was incorrect, that he remembered that Yang had responded that there was no evidence for such violation.)

When Feynman and Gell-Mann had discussed Block's question at the meeting, they realized they couldn't come up with a good empirical reason why the breaking of parity symmetry in weak kaon decays would be impossible. If the weak interaction violated parity, where else might it show up in particle physics? The weak interaction itself was not well understood. As mentioned earlier, Fermi had come up with a simple model of the prototypical weak decay, the decay of the neutron into a proton, called beta decay, but no unified picture of the different known weak decays had yet emerged.

Even as the two theoretical giants Feynman and Gell-Mann mulled over this strange possibility, and various other physicists at the conference felt strongly enough about the question to make bets, the two younger physicists Lee and Yang, both former colleagues of Gell-Mann's from Chicago, had the courage and intellectual temerity to return home and seriously explore all of the data then available to see if parity violation in the weak interaction could be ruled out. They discovered there were no experiments that could definitively answer the question. More importantly, they proposed an experiment involving beta decay itself. If parity were violated, and neutrinos were polarized so that they were made to spin in a certain direction, then parity violation would imply that electrons, one of the products of their decay, would preferentially be produced in one hemisphere compared to the other. They wrote a beautiful paper on their speculation that was published in June of 1956.

The possibility seemed crazy, but it was worth a try. They convinced their colleague at Columbia University, Chien-Shiung Wu, an expert in beta decay, to back out of a European vacation with her husband in order to perform an experiment on the decay of neutrons in cobalt 60. This was a different era from today, where the time between theoretical proposals in particle physics and their experimental verification can be decades apart. Within six months, Wu had tentative evidence not only that the electrons were emerging from her apparatus asymmetrically, but also that the asymmetry seemed about as large as was physically possible.

This convinced another Columbia experimental physicist, the future Nobel laureate Leon Lederman, whom Lee had been trying to get to conduct a similar experiment on pion decay, to perform the experiment. Again, with a speed that is almost unfathomable to physicists who grew up a generation later, Lederman and his colleague Richard Garwin reconfigured their apparatus on a Friday, after a lunchtime faculty session to discuss the possibility, and by Monday they had the result, within a day of the completion of Wu's experiment. Parity was indeed violated maximally. God, to paraphrase the baseball analogy expressed by the doubting theorist Wolfgang Pauli, was not a "weak left-hander," but a strong one.

The result produced a sensation. That so sacrosanct a symmetry of the world around us is not respected at a fundamental level by one of the four known forces of nature sent tidal waves throughout the physics world, with reverberations felt in all of the media as well. (Lee and Yang apparently had one of the first modern physics press conferences at Columbia to announce the experimental confirmation of their proposal.) In one of the quickest such developments in the history of physics, Lee and Yang shared the Nobel Prize in 1957 for their suggestion, made just one year earlier.

Why did Feynman not follow up on his question at the Rochester meeting? Once again, he had found himself close to the answer to a vital question in physics but had not pushed through to the conclusion. This tendency might reflect a character trait that would come back to haunt him: He didn't want to follow other physicists' leads. If the community was fixated on a problem, he wanted to steer clear and keep his mind open so he could work out puzzles as he liked to, from start to finish. Moreover, he hated to read the physics literature, something that was essential to the work that Lee and Yang had done.

But it might also be that Feynman had bigger fish to fry, or so he thought. Recognizing that the weak interactions violated parity was one thing, but it was not the same as coming up with a new theory of nature. This was something that Feynman yearned to do. He had long felt his work on QED was merely a technical kluge, not really fundamental, not like the equation his hero Dirac developed.

What attracted Feynman more was the possibility of coming up with a theory to unify all the different observed weak interaction phenomena, involving decays of very different types of particles, like neutrons, pions, and kaons, into a single picture. Fermi had produced a rough but beautiful, if ad hoc, model for beta decay, but the data on the weak decays of different particles was inconclusive and eluded unification. The central questions thus became: Was there a single unified weak interaction describing all of these processes, and if so, what was its form?

Feynman's sister, who was also a physicist, berated him for his timidity regarding parity nonconservation. She knew that just a little work, combined with the diligence to write it up, would have made all the difference. She urged him to not be so intellectually laissez-faire regarding his ideas about the weak interaction.

Nowadays, the violation of reflection symmetry in the weak interaction is most easily displayed by a simple statement: the exotic particles called neutrinos, the products of beta decay so-named by Fermi, which are the only known particles to interact solely by the weak interaction, are "left-handed." As I have described, most elementary particles carry angular momentum and behave as if they are spinning. Objects that spin one way would be observed to be spinning in the opposite direction if viewed in a mirror. All other known particles can be measured to behave as if they were spinning either clockwise or counterclockwise, depending on the experiment. However, neutrinos, the elusive weakly interacting particles, maximally violate mirror symmetry, at least as far as we know. They only spin one way.

Tsung-Dao Lee was actually alluding to this implication when he was describing his work with Yang at the 1957 Rochester Conference, and it grabbed Feynman's attention. Back in the early days, when Feynman was trying to first reproduce Dirac's equation as an undergraduate with Ted Welton, he had missed the boat, coming up with a simpler equation that didn't properly incorporate the spin of the electron. Dirac's equation had four different components, to describe the two different spin configurations of electrons and of their antiparticles positrons.

Feynman now realized that by using his path-integral formalism, he could naturally come up with an equation that looked like Dirac's but was simpler. It had only two components. This excited him. He recognized that if history had been different, his equation could have been discovered first, and Dirac's equation derived from it later. Of course, his equation ended up having the same consequences as Dirac's—his equation described one spin state of the electron and one for its antiparticle, and there was another similar equation that described the other two states—so it was not really new. But it did offer a new possibility. For neutrinos, which appeared to have only one spin state, his equation would, he felt, be more natural.

There was one problem, however. If one tried to mathematically incorporate this kind of equation to describe the weak interaction that led to beta decay and the production of neutrinos, it would yield results different from those that experimenters seemed to be unearthing with their experiments. The strange thing, though, was that these experimental results were inconclusive, and inconsistent with a single force at work. If one classified all the different mathematical forms one could write for the interactions of a neutron, proton, electron, and neutrino (the latter three the products of the decay of the former), one would find five different possibilities: *scalar (S), pseudoscalar (P), vector (V), axial vector (A),* and *tensor (T).* This mathematical classification scheme describes the properties of the interactions under rotations and parity flips. Each interaction has different properties. The fact that the weak interaction violates parity meant that it had to combine at least two different types of interactions, each with different parity properties. The problem was that beta-decay experiments suggested the combination was *S* and *T*, or *V* and *T*, whereas Feynman's two-component picture required a combination of *V* and *A*.

Feynman wasn't the only one thinking about these questions. Gell-Mann, who had been contemplating parity issues for some time, was also keen to unify the various weak decays, having been scooped by Lee and Yang and perhaps more bothered by that than Feynman. While Feynman was once again off in Brazil cavorting for the summer, Gell-Mann remained in California working.

But there were also others, and a particularly tragic case involves the physicist E. C. G. Sudarshan, a young Indian physicist who had come to the University of Rochester in 1955 to work with physicist Robert Marshak. In 1947 Marshak had suggested that two different kinds of mesons had been observed in experiments, one, the pion, that corresponded to the strongly interacting particle that physicists had expected, and a different one, the muon, that is now known to be simply a heavy version of the electron. Marshak was also well known in the community as the originator of the Rochester Conferences, where the central problems of the day in particle physics, like the tau-theta problem, and parity violation were hashed out.

As a graduate student Sudarshan had gained a thorough knowledge of nuclear physics, and neutron beta decay in particular, and after parity violation was discovered, he and Marshak investigated the current experimental data and decided that the conventional assignment of *S* and *T* for nuclear beta decay had to be wrong. They realized that all weak decays, including the decays of muons, could be unified together only if the interaction had a *V-A* form.

Remarkably, Sudarshan was set to present these results at the seventh Rochester Conference in 1957, where Marshak was the chair. Marshak, however, decided that as a graduate student, Sudarshan was not a delegate to the conference, and since he, Marshak, was giving a review talk on another subject, he felt he could not speak on the subject, so a prime opportunity to announce their proposal to the community was lost. As the data related to muon decay into pions, which also subsequently have weak decays into electrons and neutrinos, was not very solid, Marshak was probably hesitant to make any definitive claims at this time.

Instead, over the summer Marshak and Sudarshan completed a systematic analysis of the data up to that time, and prepared a paper proposing a *V-A* universal form of the weak interaction that Marshak would present that fall in Padua, Italy. It was a brave claim, as it required no less than four different experimental results to be wrong. During this period Marshak and Sudarshan were in California, at UCLA, and they dropped by to talk to Gell-Mann, who also set up a meeting for them with Caltech experimentalist Felix Boehm. They explained to Gell-Mann, who had been thinking about *V-A* but dropped the idea because it disagreed with some data, that they felt these experiments could be wrong. Boehm reassured them by explaining that his experimental findings were now consistent with *V-A*.

Gell-Mann, who had also realized that *V-A* was the only sensible form for a universal weak interaction, if there indeed was one, told the anxious duo of Marshak and Sudarshan that he was not planning to write a paper on the subject, but might mention the possibility in a paragraph of a long review he was writing on the weak interaction. He then headed off for a vacation, and Marshak and Sudarshan headed home.

Meanwhile Feynman had been obsessed with the idea of a universal weak interaction, perhaps his last hope for discovering a fundamental law. The confusing state of the different experiments, however, remained a stumbling block. Upon returning to Pasadena while Gell-Mann was on vacation, he learned that Boehm and Gell-Mann had been talking about the possibility of *V-A* satisfying the experimental constraints after all.

That, for Feynman, was when a bell went off. If this was true, his idea about describing neutrinos by two components in a simple mathematical form that could accommodate beta decay was right. As he later put it, "I flew out of the chair at that moment and said, 'Then I understand everything. I understand everything and I'll explain it to you tomorrow morning.' . . . They thought when I said that, I'm making a joke. . . . But I didn't make a joke. The release from the tyranny of thinking it was S and T was all I needed, because I had a theory in which if V and A were possible, V and A were right, because it was a neat thing and it was pretty." Feynman was so excited that he convinced himself he was the only person in the world who understood that *V-A* would produce a universal form for the weak interaction. Indeed, he had his own peculiar reasons for thinking so, due to, as usual, his own unique formalism. With uncharacteristic speed, he proceeded to draft a paper—his great hope for a new theory of nature, he thought.

Gell-Mann, in the meantime, returned back to Caltech to learn that Feynman was writing up his proposal, while Gell-Mann had had his own reasons for thinking of *V-A*, having to do with the symmetries of the *currents*, or the flow of charges that were associated with the particles entering and leaving the reaction. He decided, in spite of his assurance to Marshak, that he had to write his own paper.

In one of the moments that make the life of a department chairman less than fun (moments that are only too familiar from my own twelve years in such a position), the chair of the Caltech physics department decided his two stars shouldn't be playing dueling papers, and told them to team up and write a single paper. Surprisingly, Feynman and Gell-Mann agreed.

The Feynman–Gell-Mann paper was an interesting kluge of styles, but a masterpiece nevertheless. It had the best of both, Feynman's two-component neutrino formalism (which later would become useful, though at the time it seemed contrived) and Gell-Mann's brilliant thoughts on conserved quantities and symmetries associated with weak currents (which would prove useful far beyond understanding beta decay for years to come).

Needless to say, word of the Feynman–Gell-Mann paper quickly spread, and poor Sudarshan had to endure talk after talk where he heard the idea for *V-A* attributed to these two leading lights. It was true that Gell-Mann had insisted on an acknowledgment in their paper to discussions with Marshak and Sudarshan, and he always tried to write supportive letters for Sudarshan, and Feynman later acknowledged to Sudarshan that he had since been told that Sudarshan had had the idea for *V-A* before anyone else, and subsequently admitted such in public. But for years, the Feynman–Gell-Mann paper became the classic and only reference people quoted when discussing the idea.

This might have been the only time in his career that Feynman felt so driven and so excited by an idea that he wanted to publish it as his own. It was, he felt, perhaps his proudest moment, or as he put it, "There was a moment when I knew how nature worked. It had elegance and beauty. The goddamn thing was gleaming." And it was a beautiful piece of work, as one might expect from the two most creative minds in particle physics of their era. Though not earth-shattering, nor even completely surprising, and certainly not a complete theory of anything (the full theory of the weak interaction would take another decade to be written down, and another decade after that to be accepted), in spite of Feynman's subjective assessment, for the world it seemed to signify that the partnership which had started in 1954 when Gell-Mann moved to Caltech to be near Feynman had reached a kind of fruition that promised great things to come. *Time* magazine profiled the two of them among the leading lights in U.S. science: "At the blackboard the two explode with ideas like sparks flying from a grindstone, alternately slap their foreheads at each other's simplifications, quibble about the niceties of wall-length equations, charge their creative batteries by flipping paper clips at distant targets."

But this was not the beginning of a beautiful partnership. It was closer to the end. Just as their collaboration had been a forced marriage, the two geniuses, while never losing their respect for each other's abilities and ideas, carried out their future work on parallel tracks. Sure, they would consult each other for advice and periodically bounce ideas off each other. But never again would they collaborate to "twist the tail of the cosmos." Gell-Mann was soon to make his greatest contribution to physics, and Feynman was to focus on other things for almost a decade, then returning to particle physics and ironically helping to convince the world that Gell-Mann's mathematical invention, quarks, might actually be real.

CHAPTER 14

Distractions and Delights

The Prize is the pleasure of finding things out.

—Richard Feynman

At last Feynman had written up a brilliant idea in a timely manner and was satisfied—incorrectly it turned out—that he had finally been the first to unveil for the world a new law of nature. He could now revel in the pleasure of both sharing the limelight with Gell-Mann and being at the center of the physics universe.

With Gell-Mann and Feynman at Caltech, it became a place where physicists went to learn, to collaborate, or simply to reflect their own ideas off the combined brilliance of the two greatest physics minds of their generation. A succession of fertile young theorists would study there and then move on to seek out brave new worlds. Feynman himself had almost no graduate students, but the combined attraction of Feynman and Gell-Mann was enough to lure both students and postdoctoral researchers, numerous future Nobel laureates among them, to the institution.

Some were shocked at the attention they received. Barry Barish, then a budding young experimentalist from Berkeley and later a colleague of Feynman's at Caltech, gave a seminar there and was overwhelmed to see Feynman and later to be peppered with questions by him. Barish recently related to me how self-satisfied and important he felt at the time. That is, until others told him that Feynman attended all of the seminars and was full of questions—there was nothing special about his visit.

At the same time the place could be intimidating. Gell-Mann could be cutting, usually in private. On the rare occasions when Feynman thought little of one's work, he could be openly dismissive or worse. What would set him off would not always be clear. Certainly he had little patience for nonsense, but he also clearly reacted negatively to approaches that were valid, perhaps even brilliant, but reflected a style he didn't like. An example is the reception a young theorist, Steven Weinberg, received when he went to Caltech to present a seminar on his ideas. Weinberg, who later became one of the world's most respected and accomplished physicists (and later shared the Nobel Prize for coming up with a full theory of the weak interaction, unifying it with electromagnetism), often sought out detailed formal solutions, working from the general to the particular—the opposite of how Feynman often worked. This physicist of such obvious substance was so mercilessly questioned by both Feynman and Gell-Mann that he almost could not finish his talk.

Feynman's wrath was normally restricted to those who he felt were abusing physics by making unfounded claims, usually on the basis of insufficient evidence. To Feynman, the physics came first, and it didn't matter who the culprit was. Perhaps the most famous example that I am personally aware of involved a future Nobel laureate, Fred Reines, who in 1995 won the prize for a 1956 experiment that first verified the existence of neutrinos. Reines had continued his work on neutrinos coming from nuclear reactors, and much later, in 1975 claimed to have evidence that neutrinos, which come in several types, were oscillating from one type into another as they traveled outward from the reactor. If true, this result would have been hugely significant (it turns out that neutrinos do oscillate—just not in the way Reines had claimed). Feynman examined the data and demonstrated that the claimed effect was not substantiated and publicly confronted Reines with his results, ultimately putting to rest this false positive. This embarrassment might have contributed to an almost forty-year delay in awarding Reines the Nobel Prize for his original discovery.

In any case, back in 1957 Feynman's work with Gell-Mann on the weak interaction released him from a burden he had carried with him over the years as his fame in the community had continued to increase even as he remained personally skeptical about the significance of his work on QED. While he never lost interest in particle physics, he seemed freer to wander further afield and try his talents elsewhere.

At the same time, his mind was also wandering over other domains. His personal relationships were getting messy again. In 1958, in Geneva to present an overview of weak interaction physics at one of the first Rochester Conferences held abroad, he had planned to travel with the wife of a research associate at Caltech, with whom he had apparently been having an affair. This was happening even as another affair was coming to a brutal end, also with a married woman, and one whose husband was prepared to sue for damages. Ultimately that uproar died down, and the spurned lover eventually returned to Feynman both a gold medal he had been awarded as a part of the Einstein Prize in 1954 and some drawings by Arline.

In Geneva, Feynman found himself alone, since his lover had decided she would avoid Switzerland and meet him later in England. On the beach he met a young twenty-four-year-old Englishwoman, Gweneth Howarth, who was traveling around the world as an au pair, and at the time was in possession of no less than two other boyfriends, so in that sense they were even. Not surprisingly, he took her to a club that night, but much more surprisingly, before he left Geneva he invited her to come work for him as his maid in the United States, and he offered to assist with the necessary immigration procedures. (Whether he went on to meet with his other lover in London is not recorded.) There were, of course, inevitable delays, and even as Feynman continued to deal with the consequences of his other spurned lover, with whom he had considered marriage, Gweneth got involved in several romantic liaisons and periodically changed her mind about coming. Thus, while there had clearly been a spark between them, it is hard to know exactly what Feynman, or Gweneth, had in mind with all of this.

It was brought to Feynman's attention that given the circumstances it might seem inappropriate at best, or illegal at worst, for a forty-year-old man to help transport this twenty-four-year-old woman to live under his roof, so his colleague, a delightful and free-spirited experimental physicist named Matthew Sands, arranged the paperwork in his name. Finally, after more than a year of delays, Gweneth arrived in Pasadena in the summer of 1959, and helped convert the house of this clearly lonely bachelor into a home.

While she dated other men—whether a pretense or not, it is hard to say—this behavior too declined, and she would eventually accompany Feynman to social events, often leaving separately to keep up appearances. A little over a year after her arrival, Feynman asked her to marry him.

This story was the stuff of B-rated movies, not real life, and there was every reason to suggest that Feynman's rash behavior would end up, like so many of his other romantic escapades, in disaster. But it didn't. Two years later they had a son, Carl, and a dog. Richard's mother moved out to live nearby, and he purchased a home close to his colleague and collaborator Gell-Mann and his new British bride. Feynman had become a family man. He and Gweneth later adopted another child, their daughter Michelle, and remained happily married until his death.

Feynman's personal life finally settled down—it would after all have been impossible to have become more unsettled than it had been—but his mind remained restless. He considered moving into another field and toyed with genetics, egged on by his friend, physicist-turned-biologist and future Nobel laureate Max Delbrück. But that didn't take for long.

He continued to work with Gell-Mann on the weak interaction and to tease him at seminars (though the sparring perhaps became a little more pointed over time), but his heart didn't seem to be in this work. The two of them had an idea that there might be two different kinds of neutrinos in order to explain a puzzling experimental result, but Feynman lost interest in it and refused to write it up. Subsequently Leon Lederman, Melvin Schwartz, and Jack Steinberger won the Nobel Prize for experimentally verifying that this was indeed the case. On another paper, with Gell-Mann and several other colleagues, Feynman agreed to collaborate, and then after the preprint was sent out, he begged, successfully, to have his name removed before publication.

In 1961 an unusual opportunity came along that opened up his creative energy in a totally new way and helped catapult Feynman to a new rank within the physics community and beyond. It did not involve discovering a new law of nature, but rather discovering new ways to teach physics.

Undergraduates at Caltech were required to take two years of introductory physics courses, and like most such courses they were a disappointment, especially for the best and brightest students who had been excited by physics in high school and wanted to learn about relativity and the modern wonders and didn't want to start over studying balls rolling down inclined planes. At the instigation of Matthew Sands, who had been discussing the idea with Feynman for some time, the physics department, and eventually the chair, Robert Bacher, the same one who had enforced the shotgun marriage of Feynman and Gell-Mann, decided to revamp the course. Again, at Sands's suggestion, it was agreed that Feynman would take one entering class through the entire introductory sequence. Even though Feynman was not widely recognized for his teaching, his teaching reviews at Cornell had been very good, and he was well known in the community as a uniquely gifted expositor when he put his mind to it. His remarkable energy, his colloquial manner, his physics intuition, his Long Island accent, and his inherent brilliance gave him a riveting aura behind any podium.

Feynman took up the challenge and then some. He had

tirelessly devoted his whole life to rebuilding in his own mind the entire edifice of physical law. The adventure of personal understanding had driven him since his childhood. Now was an opportunity for him to put this picture out there for others. (I have since discovered, while reading through Matthew Sands's memoirs, that he used almost this same language to convince Feynman to teach the class.) He could put his own brand, not just on physics at the forefront, but on the very basic ideas that form the heart of our physical understanding. Over the next two years he devoted more intense energy and creativity into developing his lectures than he had put into anything since the war.

The timing was perfect. This expenditure was possible at that time in part because his wanderlust had subsided. With the stability of his marriage and domestic life, he was able to focus less on his own needs and desires, had fewer motivations to seek out adventures to mask his loneliness, and more importantly, could settle down in one place for the time required to sketch out a completely new introduction to the fundamentals of physics. He could show others not only how he personally understood the world, but also what had excited him enough to learn about it. He could make new connections, which is what science is all about, in unraveling the mysteries of the physical universe. He wanted to quickly take students to the exciting forefront mysteries, but at the same time show that they were not all esoteric, that many were connected to real phenomena as immediate as boiling oatmeal, or predicting the weather, or the behavior of water flowing down a tube.

Every day he arrived at class before the students, smiling and ready to regale them with yet another totally original presentation of everything from classical mechanics to electromagnetism, gravity, fluids, gases, chemistry, and ultimately even quantum mechanics. He would march back and forth behind a huge demonstration table and in front of a mammoth blackboard, yelling and grimacing and cajoling and joking. And by the end of the class he would make sure not only that the entire chalkboard was full, but that he had completed the circle of ideas he had set out at the beginning of class to discuss. And he wanted to show students that their lack of knowledge didn't need to compromise their understanding, that with hard work even freshmen could address, in exact detail, some modern phenomena.

Most of all, he wanted to present a guide for understanding or, as he almost called it, "a guide for the perplexed"—after the title of the famous tract by the eleventh-century philosopher Moses Maimonides. As he said,

> I thought to address them [the lectures] to the most intelligent in the class and to make sure, if possible, that even the most intelligent student was unable to completely encompass everything that was in the lectures—by putting in suggestions of applications of the ideas and concepts in various directions outside the main line of attack. For this reason, though, I tried very hard to make all the statements as accurate as possible, to point out in every case where the equations and ideas fitted into the body of physics, and how—when they learned more—things would be modified. I also felt that for such students it is important to indicate what it is that they should—if they are sufficiently clever—be able to understand by deduc-

> tion from what has been said before, and what is being put in as something new.

In his excitement he also wanted to connect physics with the rest of science, to show that it was not an isolated island. He introduced the physiology of color vision and the very mechanical engineering applications that had so interested him when he was a student, and of course he described his own discoveries as well.

The department realized something special was happening, and Feynman was given great support and encouragement. Every week he would meet with other faculty who were assigned, under the supervision of Matthew Sands and Robert Leighton, to devise problem sessions and extra recitations to help the students. Since Feynman wasn't teaching out of any textbook, it was necessary to have these meetings, and these instructors and assistants had to work full-time, both to keep up and to develop appropriate teaching materials to complement the course.

Soon, word of what was going on in the large lecture hall at Caltech spread, and graduate students and faculty began to trickle in to listen, even as the terrified and overwhelmed undergraduates stopped coming. As perhaps might have happened only at a school like Caltech, the department urged him to continue lecturing for a second year, in spite of the fact that many of the students couldn't pass his exams.

The lectures were also recorded so that Sands and other colleagues, chiefly Leighton, could transcribe and edit them. Ultimately a three-volume set of "red books" appeared for sale around the world. Never before in modern times had someone so comprehensively or so personally re-created from scratch and reorganized the entire knowledge base and presentation of the basic principles of physics. This was reflected in the name given to the set: *The Feynman Lectures on Physics.*

This is significant. It ascribed a unique status to a single scientist, and I don't know of anyone else, in physics at least, for whom such a title would have even been considered. Feynman was in the process of becoming a physics icon, and the title was testimony not only to the nature of the material but also to the special place which Feynman was coming to occupy in the physics world.

In the end the actual course was a mixed success. Few of the students, even Caltech students, could follow all of the material. However, over time those who were lucky enough to attend the course had memories that mellowed. Many of the former students remembered it as the experience of a lifetime, echoing the words of Nobel laureate Douglas Osheroff, who later said, "The two-year sequence was an extremely important part of my education. Although I cannot say that I understood it all, I think it contributed most to my physical intuition."

But while the undergraduate guinea pigs at Caltech may have suffered (though Sands disputed this notion and argued that most of the students kept up at some level), the *Feynman Lectures* became a staple for anyone who planned to become a physicist. I remember buying my own copy when I was an undergraduate, and I would read tidbits at a time, wondering whether I would ever really grasp all the material, and hoping that one of my professors would use the book. Perhaps luckily for me, none of them did. Most who tried, found the experiment to be disappointing. The material was too demanding for the average physics class, and too revolutionary.

Nevertheless, the books remain in print, a new revised set appeared in 2005, and every year, a new crop of students buys them, opens them up, and begins to experience a whole new world.

Unfortunately, for Feynman, Sands, and Leighton, however, all royalties continue to go to Caltech (although Feynman's family later sued Caltech over rights to one of the lectures that Caltech packaged as a book and audiotape). Later on, Feynman commiserated with friend, physicist, and author Philip Morrison, after being called a physics giant, "Are we physics giants, business dwarfs?"

The experience of teaching this course coincided with a general outburst of popular activity by Feynman, whose charismatic style was beginning to make waves well beyond the confines of the physics community. Already in 1958 he had agreed to be an advisor of a television program that Warner Brothers was producing, and in a letter regarding that production he indicated both his experience with popular outreach and his philosophy: "The idea that movie people know how to present this stuff, because they are entertainment-wise and the scientists aren't is wrong. They have no experience in explaining ideas, witness all movies, and I do. I am a successful lecturer in physics for popular audiences. The real entertainment gimmick is the excitement, drama, and mystery of the subject matter. People love to learn something, they are 'entertained' enormously by being allowed to understand a little bit of something they never understood before."

Around that time he also participated in what I believe

was his first television interview, which aired shortly before Gweneth arrived in the United States. He was clearly excited about being on television, and advised her, "If you came 2 weeks earlier I'd sure have a lot for you to do—I'm going to be on television, in an interview with a news commentator on June 7th and there may be a lot of letters to answer." The interview was a masterpiece, far exceeding the quality and intellectual depth of interviews performed nowadays, but because there was a frank discussion of religion in it, the network decided to air it at a different time than advertised, so the viewing audience was smaller.

The television production he had become an advisor on, called *About Time*, finally aired on NBC in 1962. It elicited a large reaction among viewers and began to further establish Feynman's popular credentials. His prowess as a lecturer to lay audiences led to an invitation to give the prestigious Messenger Lectures at Cornell. This set of six lectures became famous and were compiled into a wonderful book titled *The Character of Physical Law*. (This was the very book that my summer school physics instructor had recommended I read in order to get me more excited about physics.) The lectures were also recorded on film, and recently Bill Gates bought the rights to them so they could be available online. (Gates said that if he had had access to them when he was a student, before he dropped out of Harvard, his life might have changed.) They, more than any other recorded image or document, capture the real Feynman, playful, brilliant, excited, charismatic, energetic, and no nonsense.

Finally, on October 21, 1965, Feynman was "canonized," cementing his status forever among scientific and lay audiences. Feynman, along with Sin-Itiro Tomonaga and Julian

Schwinger, shared the 1965 Nobel Prize for their "fundamental work in quantum electrodynamics with deep ploughing consequences for the physics of elementary particles." Like that of every other Nobel laureate, Feynman's life was forever changed, and he worried about this effect. While there is little doubt he enjoyed celebrity, he didn't like pomp and circumstance, and motivated from the attitude he had gleaned from his father as a child, he truly distrusted honorific titles. He matched his thoughts with actions. He had decided as a young man, graduating from MIT, that honorary degrees were silly—those being honored with the degrees had not done as much work as he had to attain his—so he refused to accept any that were offered to him. In the 1950s he had been elected to membership in the prestigious National Academy of Sciences, for many scientists the highest recognition they can get from their colleagues. Beginning in 1960, Feynman began a long and involved process to resign from the National Academy because he viewed its prime purpose as to determine who could be "in" and who was not "worthy." (In a famous episode years later, Carl Sagan was rejected for membership, many think at least in part because of his popularization efforts.) He stopped listing it among his honors (he asked the NBC television people, for example, to remove it from his biographical sketch for the 1962 television program), but it took ten years before the Academy officials finally made his resignation official.

It is hard to know how serious Feynman was, but he later wrote that for a moment he considered refusing the Nobel Prize for these same reasons—who cared if someone in the Swedish Academy decided that his work was "noble" enough. As he famously said, "The prize is the pleasure of finding things out." But he quickly realized that this would generate more publicity than getting the prize, and might lead to the impression that he thought he was "too good" for it. He claimed that what the Nobel Committee should do instead is quietly let the prize winners know in advance of its decision, and give them time to back out quietly. According to Feynman, he wasn't the only one who had this idea. His idol Dirac had felt the same way.

In spite of his misgivings, Feynman clearly felt some validation by the prize, and as his former student Albert Hibbs said, he probably would have felt worse if he hadn't gotten it. He also liked the fame the prize and other recognitions brought, not least because it gave him more freedom to act as he wished.

Be that as it may, in spite of his intense nervousness about messing up during the official ceremony, about bowing and wearing formal attire, and about walking backward in the presence of the Swedish king, Feynman persevered, attended the ceremony, and prepared a beautiful Nobel lecture giving a truly personal history of his journey to discover how to tame the infinities of QED. And even as late as 1965, Feynman still felt the program of renormalization that he had spearheaded was merely a way of sweeping problems of infinities under the rug, not curing them.

Associated with the awarding of the prize was the question that often comes about because of the phrasing of Mr. Nobel's will, which seems to imply that only one person can win. It was clear in this case that Julian Schwinger, Freeman Dyson, and Sin-Itiro Tomonaga all deserved a share of the prize, but why didn't Dyson receive it? He had so skillfully demonstrated the equivalence of the seemingly totally different methods of deriving a sensible form of QED, and had followed that up by essentially providing a guide to teach the rest of the physics world how to do the appropriate calculations. Dyson, you will recall, was also essentially the one whose papers advertised Feynman's results before Feynman ever wrote his own papers, and who ultimately helped explain to the world that Feynman's methods were not ad hoc, but as well grounded and much more physically intuitive and calculationally simple than the other approaches. It was thus Dyson who had helped the rest of the world understand QED, while establishing Feynman's methods as the ones that would ultimately take root and grow.

If Dyson had bad feelings about not receiving the prize, he never verbalized it. Quite the opposite in fact. As he later put it, "Feynman made the big discoveries, and I was just really a publicizer. I got well rewarded for my part in the business—I got a beautiful job here at the Institute, set up for life, so I've nothing to complain about! No, I think that it was entirely right and proper. Feynman's was one of the best-earned Nobel Prizes there ever was, I would say."

CHAPTER 15

Twisting the Tail of the Cosmos

I think I've got the right idea, to do
crazy things . . .

—Richard Feynman

The years between 1957 and 1965 represented a transitional period in Richard Feynman's life. Personally, he went from womanizer to family man, from solitary wanderer to domesticated husband and father (though he never stopped seeking adventures, now, more often than not, with his family). Professionally, he went from someone urgently working, essentially for his own immediate pleasure, at the leading edge of physics to someone who had begun to give back to the world the wisdom he had gained from his years of thinking about nature (though he probably would never have claimed that what he had was wisdom).

In the meantime he had become one of the most well-known and colorful expositors and teachers in physics, and, in part, its conscience. He remained acutely aware, and ensured whenever he could that his colleagues and the public at large did not lose track of what science was and what it wasn't, what excitement could be gained from studying it, and what nonsense could result from over-interpreting its signals, from unfounded claims, or simply by losing touch with it completely. He felt strongly that science required a certain intellectual honesty, and that the world would be a better place if this was more widely understood and practiced.

This is not to suggest that Feynman the person changed in any fundamental way. He remained intensely interested in all aspects of physics and, as I just mentioned, he continued to seek out adventures, just adventures of a different kind. Besides exotic trips with his wife, he took up two activities that might be called sublimation. One involved working on his calculations almost every day in a Pasadena strip joint, where he could watch the girls whenever calculations were going badly, and another involved combining a long-standing interest in drawing with a long-standing interest in nude women that he could draw. He actually became fairly accomplished at this, which is paradoxical since as a young man he had scoffed at music and art but, by middle age had taken up both. Equally paradoxical was a new fascination, in the 1960s, with visiting various New Age establishments like Esalen, where he would enjoy both the scenery and the opportunity to relieve the participants of their beliefs in "hokey-pokey" fairy tales, as he put it. Perhaps the attraction of the nude baths, and the fascination with interacting with a completely different sort of individual than he would otherwise, outweighed his own lack of tolerance for those who abused the concepts of physics, like quantum mechanics, to justify their "anything goes" mentality.

In his professional life, as his fame increased, he moved to aggressively protect his time. He wanted to ensure that he didn't become a "great man" in the traditional sense, encumbered by a host of administrative responsibilities, which he shirked at every possible moment. He even had recorded a bet with Victor Weisskopf, when he visited CERN in Geneva after he received the Nobel Prize, that within ten years he would not hold a "responsible position"—that is, a position that "by reasons of its nature, compels the holder to issue instructions to other persons to carry out certain acts, notwithstanding the fact that the holder has no understanding whatsoever of that which he is instructing the aforesaid persons to accomplish." Needless to say, he won the bet.

His growing fame encouraged another tendency that had paid off for him in the past, though in the long run it cost him what could have been a great deal more success in continuing to lead in discovering new physics. He became more and more convinced that in order to blaze new paths, he needed to disregard much of what others were doing, and in particular not focus on the "problems du jour."

Physics is, after all, a human social activity, and at any time there is often consensus about what the "hot" problems are, and which directions are most likely to lead to new insights. Some view this faddishness as a problem, as for instance, the fascination of much of the community over the past twenty-five years with string theory, a mathematically fascinating set of ideas whose lack of direct contact with the empirical world has been outweighed only by the increasing confusion about what it might predict about nature. (Nevertheless, in spite of this, the mathematics of string theory has led to a number of interesting insights about how to perform calculations and interpret the results of more conventional physics.)

It is inevitable that groups of people with similar interests will get excited about similar things. And ultimately, fads in science don't matter, because first, all of the activity inevitably reveals the warts as well as the beauty marks quicker than would otherwise have been the case, and second, as soon as nature points out the right direction, scientists will jump off a sinking ship faster than rats in a storm.

In order for science to be healthy, it is important that not *all* scientists jump on the same bandwagon, and this was the point that Feynman focused on, almost to obsession. He was so talented and so versatile that he was able, if necessary, to reinvent almost any wheel and usually improve it in the process. But by the same token, reinventing the wheel takes time and is rarely worth the trouble.

It wasn't just that he could take this road; it was that he often felt he had to. This was both a strength and a weakness. He really didn't trust any idea unless he had worked it out from first principles using his own methods. This meant that he understood a plethora of concepts more deeply and thoroughly than most others, and that he had a remarkable bag of tricks from which he could pull magic solutions to a host of varied problems. However, it also meant that he was not aware of brilliant developments by others that could have illuminated his own work in new ways, leading him further than he could have gotten on his own.

As Sidney Coleman, a brilliant and remarkably well-respected Harvard physicist who had been a student of Gell-Mann's at Caltech in the 1950s and who had interacted with Feynman throughout his career, put it, "I'm sure Dick thought of that as a virtue, as noble. I don't think it's so. I think it's kidding yourself. Those other guys are not all a collection of yo-yos. . . . I know people who are in fact very original and not cranky but have not done as good physics as they could have done because they were more concerned at a certain juncture with being original than with being right. Dick could get away with a lot because he was so goddamn smart."

Feynman did get away with a lot. But could he have done much more if he had agreed every now and then to build on well-trodden paths rather than seek out new ones? We will never know. However, an honest assessment of his contributions to science from 1960 or so onward demonstrates several trends that continued to repeat themselves. He would explore a new area, developing a set of remarkably original mathematical techniques and physical insights. These would ultimately contribute to central developments by others, which would lead to a host of major discoveries and essentially drive almost every area of modern theoretical and experimental physics. This ranged from his work in condensed matter physics to our understanding of the weak and strong interactions, to the basis of current work in quantum gravity and quantum computing. But he himself did not make the discoveries or win prizes. In this sense, he continued to push physics forward as few modern scientists have, opening up new areas of study, producing key insights, and creating interest where there had been none before, but he tended to lead from the rear or, at best, from a side flank.

Whether or not this would have disturbed him is unclear. In spite of his natural tendency to showboat, as I have described, he was ultimately more interested in being right than being original, and if his work led others to uncover new truths, he might remain skeptical of their results for a long time, but eventually the satisfaction of having provided illumination in the darkness gave him deep pleasure. And by concentrating on difficult problems the mainstream would not approach, he increased his chances of providing such illumination.

FEYNMAN'S FIRST FORAY well off the beaten path involved his desire, beginning around 1960, to understand how one might formulate a quantum theory of gravity. There were good reasons for his interest. First, while developing such a theory had thus far eluded all who had thought about it, he had already been successful in developing a consistent quantum theory of electromagnetism when others had been stymied, and he thought his experience with QED might lead somewhere useful. Second, Einstein's general relativity had long been considered the greatest scientific development since Newton. It was, after all, a new theory of gravity. But when one considered its behavior on small scales, it appeared to be flawed. The first person who could set this theory straight would surely be viewed as the rightful heir to Einstein. But perhaps the biggest attraction for Feynman was that no one else, at least no one who really mattered, was thinking about the problem. As he put it, in a letter to his wife from a conference on gravity that he attended, in Warsaw in 1962, "This field (because there are no experiments) is not an active one, so few of the best men are doing work in it."

That was probably somewhat of an overstatement, but in truth the study of general relativity had become a field unto itself since Einstein's great discovery of his classical field equations in 1915. Because general relativity implies that matter and energy affect the very nature of space itself, allowing it to curve, expand, and contract, and that this configuration of space then affects the subsequent evolution of matter and energy, which then continues to impact on space, and so on, the theory is both mathematically and physically far more complicated than Newton's theory of gravity had been.

A great deal of work was done to find mathematical solutions to these equations in order to explain phenomena ranging from the dynamics of the universe to the behavior of the last moments of stars as they burn out their nuclear fuel. The equations were complicated enough, and their physical interpretation confusing enough, that tremendous ingenuity and mathematical prowess were required, and a small industry of experts had developed to investigate new techniques to deal with these issues.

To get a sense of how complicated the situation actually was, it took a full twenty years and lots of detours down blind alleys and errors, including some famous ones by Einstein himself, before scientists realized that general relativity was incompatible with a static and eternal universe, which was the preferred scientific picture of the cosmos at the time. In order to allow for such a universe in which our galaxy was surrounded by static empty space, Einstein added his famous cosmological constant (which he later called his biggest blunder).

The Russian physicist Alexander Friedmann first wrote down the equations for an expanding universe in 1924, but for some reason the physics community largely overlooked them. The Belgian priest and physicist Georges Lemaître independently rediscovered the equations and published them in an obscure journal in 1927. While Lemaître's work did not receive general notice, Einstein certainly was aware of it, and wrote to Lemaître: "Your calculations are correct, but your physics is abominable."

It was not until the 1930s, after Edwin Hubble's observation of the expanding universe through the motion of distant galaxies, that Lemaître's work was translated to English and began to receive general acceptance, including by Einstein. In 1931 Lemaître published his famous article in *Nature* outlining his "primeval atom" model, which eventually became known as the *big bang*. Finally, in 1935, Howard Robertson and Arthur Walker rigorously proved that the only uniform and isotropic space (by then it had become recognized that our galaxy was not alone in the universe, and that space was largely uniform in all directions with galaxies everywhere—an estimated 400 billion in our observable universe) was the expanding big bang described by Friedmann and Lemaître. After that, the big bang became the preferred theoretical cosmological model, although it actually took another thirty years—after Feynman began his work—before the actual physical signatures resulting from a Big Bang were seriously explored, and the discovery of the cosmic microwave background radiation put it on an unequivocal empirical footing.

While it took two decades to sort out the cosmological implications of general relativity, the most familiar of all situations associated with gravity, the gravitational collapse of a spherical shell of matter, remained confused for far longer, and is still not fully understood.

Within a few months of Einstein's development of general relativity, the German physicist Karl Schwarzschild wrote down the exact and correct solution describing the nature of space and the resulting gravitational field outside of a spherical mass distribution. However, the equations produced infinite results at a finite radius from the center of the distribution. This radius is now called the *Schwarzschild radius.* At the time, it was not understood what this infinity meant, whether it was simply a mathematical artifact or reflected some new physical phenomena taking place at this scale.

The famous (and eventually Nobel Prize–winning) Indian scientist Subrahmanyan Chandrasekhar considered the collapse of real objects like stars and argued that for stars larger than about 1.44 times the mass of our sun, no known force could stop their collapse down to this radius or smaller. The famous astrophysicist Arthur Eddington, however, whose own observations of a total eclipse in 1919 had provided the first experimental validation of general relativity (and catapulted Einstein to world fame), violently disagreed with this result and ridiculed it.

Ultimately Robert Oppenheimer demonstrated that Chandrasekhar's result was correct after a slight refinement (about 3 solar masses or greater). But the question still remained, What happened when stars collapsed down to the Schwarzschild radius? One of the strangest apparent implications of the theory was that from the point of view of an outside observer, as massive objects collapsed, time would appear to slow down as the Schwarzschild radius was approached, so that the objects would seem to "freeze" at this point before they could collapse further, leading to the name *frozen stars* for such objects.

For these reasons and others, most physicists believed that collapse inside the Schwarzschild radius was physically impossible, that somehow the laws of physics would naturally stop the collapse before the Schwarzschild radius was reached. However, by 1958 scientists understood that the apparent infinities associated with the Schwarzschild radius were mathematical artifacts of the coordinate system used to describe this solution, and that nothing unphysical would happen as objects traversed this radius, now called the *event horizon* because once inside, objects could no longer communicate to the outside world.

While nothing untoward might happen at the event horizon, in 1963 Roger Penrose demonstrated that anything that falls through the event horizon would be doomed to collapse to an infinitely dense "singularity" at the center of the system. Once again, dreaded infinities were cropping up, this time not just in calculations of the interactions of particles, but in the nature of space itself. It was speculated, though it has not been proved, and indeed several tentative numerical counterexamples have been suggested recently, that all such singularities are shielded from outside observers by an event horizon and therefore cannot be seen directly. If true, this would have the effect of sweeping under the rug the problem of what actually happens inside such an object, but it would clearly not resolve the key physical question of whether such singularities exist.

In 1967 John Wheeler, Feynman's former supervisor, who had earlier argued most strongly that collapse inside the Schwarzschild radius would be impossible in a sensible universe, gave in to the possibility and forever enshrined such collapsed objects with the enticing name *black holes*. Whether it is their name that has provoked such interest, black holes remain at the center of all modern controversies concerning our understanding of gravity at small scales and strong fields.

These issues surrounding the interpretation of classical solutions of Einstein's equations and the nature of gravitational collapse were the focus of activity in the community of theorists studying gravity when Feynman began to turn his attention to this field. What was most striking, perhaps, was how the study of gravity had evolved to become almost a separate and isolated field of physics. After all, Einstein had seemingly demonstrated that gravity was completely different from all of the other forces of nature. It resulted from the curvature of space itself, whereas the other forces seemed to operate quite differently—based on the exchange of elementary particles moving through space, for example. Even textbooks tended to treat general relativity as an entirely self-contained field that could be understood apart from almost all of the rest of physics.

Feynman, however, rightly believed that such a separation was artificial. At small scales, quantum mechanics reigned, and ultimately if one was going to attempt to understand gravity at small scales, one would need to use the tools that Feynman and others had developed to understand how classical theories like electromagnetism—which on the surface seems similar, being a long-range force that falls off with the square of distance out to infinity—could be made consistent with the principles of quantum mechanics. Perhaps by approaching gravity, then, as he and others had approached QED, they might be able to gain valuable new insights.

Feynman began thinking about these issues seriously in the mid-1950s, shortly after he had finished his own work on QED, and had discussed them with Gell-Mann during Christmas of 1954, by which time he had already made great progress. However, it was not until 1962–63 that he completed and formalized his thoughts, during a year-long graduate course he taught at Caltech. His lectures on the subject were turned into a book much later, released for popular consumption in 1995 and not surprisingly titled *The Feynman Lectures on Gravitation.* This title was especially fitting because he taught this graduate course at the same time that he was developing and teaching the second year of his famous introductory course on which the more well-known *Feynman Lectures* was based. It is no wonder that he was exhausted at the end of this period.

He explained the motivation for his approach in a 1963 scientific paper that summarized his results, and apologized for considering the quantum aspects of gravity, which were then, as they are now, far removed from any possible experimental verification: "My interest in it [the quantum theory of gravitation] is primarily in the relation of one part of nature to another. . . . I am limiting myself to not discussing the questions of quantum geometry. . . . I am not trying to discuss any problems which we don't already have in present quantum field theory of other fields."

It is difficult, in the current climate, where such great interest has developed in unifying the different forces of nature, to realize how revolutionary Feynman's approach was. The idea that gravity might not be so special or self-contained, was almost heretical, especially to the closed community of scientists who treated it as a special jewel, to be worked with special tools not available to ordinary physicists. As might be expected, Feynman had little patience for such an effete viewpoint; it flew in the face of all of his beliefs about science. While at the second conference on gravity that he attended, in Warsaw (the first, in Chapel Hill in 1957, was presumably more enjoyable), he wrote to Gweneth:

> I am not getting anything out of this meeting . . . there are hosts (126) of dopes here—and it is not good for my blood pressure—such inane things are said and seriously discussed—and I get into arguments outside of the formal sessions . . . whenever anyone asks me a question or starts to tell me about his "work." It is always either—(1) completely un-understandable, or (2) vague and indefinite, or (3) something correct that is obvious and self-evident worked out by a long and difficult analysis and presented as an important discovery, or (4) a claim, based on the stupidity of the author that some obvious and correct thing accepted and checked for years is, in fact, false (these are the worst—no argument will convince the idiot), (5) an attempt to do something probably impossible, but certainly of no utility, which, it is finally revealed, at the end, fails, or (6) just plain wrong. There is a great deal of "activity in the field" these days—but this "activity" is mainly in showing that the previous "activity" of somebody else resulted in an error or in nothing useful or in something promising. . . . Remind me not to come to any more gravity conferences.

Feynman began by arguing that gravity was even weaker than electromagnetism, and therefore—just as one could try to understand the quantum theory of the latter by considering first the classical theory, and then adding small quantum corrections order by order—the same procedure should work for gravity. Hence, it was worth investigating whether the infinities that resulted when one went beyond the lowest-order approximation in electromagnetism also appeared in gravity, and whether one might remove them in the same way as one had done in QED, or whether new complications might result that could give insight into the nature of gravity itself.

In electromagnetism, forces result from the interaction of charged particles and electromagnetic fields, the quanta of which are called photons. Remarkably, as far as I can determine, Feynman was the first to suggest that one might treat quantum gravity just like any other quantum theory, and in particular like the quantum theory of electromagnetism, which on the surface has a great deal of similarity to gravity. To do this he explored a remarkable idea: Let's say Einstein had not come up with general relativity. Could someone have instead derived Einstein's equations just by thinking about the classical limit of quantum particles interacting with quantum fields? While Feynman was not the first to explore such a possibility or to draw a positive conclusion in this regard—in fact, Steven Weinberg performed the most general and powerful exploration of this question in 1964, and elaborated in his beautiful text on gravity and cosmology in 1972, and again in a later paper in 1979—Feynman's original analysis created the modern mindset for the more recent reappraisals of the theory.

The claim is remarkable: Forget all about geometry and the fascinating notions about space and time that seem to be at the basis of general relativity. If one considers the exchange of a massless particle (just as a photon is a massless particle that conveys the electromagnetic force), then if the massless particle in question has quantized spin 2 instead of spin 1 as a photon does, the only self-consistent theory that results will, in the classical limit, essentially be Einstein's general relativity.

This is a truly amazing claim because it suggests that general relativity is not that different from the theories describing the other forces in nature. It can be described by the exchange of fundamental particles just like the rest. All the geometric baggage comes out after the fact, for free. In fact, there are subtleties in the actual true statement of the claim, coming from what is meant by "self-consistent," but these are really just subtleties. And Weinberg, as I have indicated, was able to prove a more general version of this claim, relying simply on the properties of the interactions of a massless spin 2 particle and the symmetries of space that arise in special relativity.

But these subtleties aside, this new picture of gravity and general relativity created a completely novel bridge between general relativity and the rest of physics that was not there before. It suggested, just as Feynman had hoped it would, that one might use the tools of quantum field theory not only to understand general relativity but also to unify it with the other forces in nature.

First, what are these massless spin 2 particles and what do they correspond to in nature? Well, recall that photons, the quanta of the electromagnetic field, are just quantized versions of classical electromagnetic waves, the waves of electric and magnetic fields that James Clerk Maxwell first showed result from jiggling an electric charge, which is the source of the electromagnetic field. These fields we experience with our eyes as light, our skin as heat from the sun, as radio waves with our radios, or microwaves with our cell phones.

Einstein had shown, shortly after he developed general relativity, that mass, which is the source of gravity, could produce a similar effect. If a mass is moved in just the right way, a new type of wave will be emitted—a gravitational wave, which is literally a wave in which space compresses and expands along the wave, and will travel out at the speed of light, just as photons do. In 1957, when Feynman first discussed his ideas at a physics meeting, many in the audience were dubious that gravitational waves even existed. (In fact, Einstein himself was earlier deterred by H. P. Robertson from publishing a paper denying their existence.) However, in 1993 Joseph Taylor and his former student Russell Hulse received the Nobel Prize for convincingly demonstrating that a pair of orbiting neutron stars was losing energy at the exact rate predicted by general relativity for the emission of gravitational waves from this system. While scientists have yet to directly detect gravitational waves, because gravity is so weak, large terrestrial experiments have been designed to do so, and plans are underway to build a very sensitive detector in space.

Gravitational waves are emitted only from objects in which the distribution of mass is changing in a nonspherically symmetric way. Physicists call the kind of radiation emitted by such a distribution *quadrupole* radiation. If one wanted to encode this kind of directional anisotropy by associating particles with the emitted waves, these primary "quanta" would have to have a spin 2, which is precisely why Feynman first explored this option. The quantum of gravitational waves is called a *graviton*, in analogy to a photon.

Having demonstrated that gravity can result simply from the exchange of gravitons between masses, just as electric and magnetic forces result from the exchange of photons between charges, Feynman then proceeded to use precisely the kind of analysis that had stood him in such good stead with QED to calculate quantum corrections to gravitational processes. The effort was not so simple however. General relativity is a far more complex theory than QED because while photons interact with charges in QED, they do not interact directly with each other. However, because gravitons interact with *any* distribution of mass or energy, and since gravitons carry energy, gravitons interact with other gravitons as well. This additional complexity changes almost everything, or at least makes almost everything harder to calculate.

Needless to say, Feynman did not find that a consistent quantum theory of gravity interacting with matter, without any nasty infinities, could be derived by simply treating general relativity as he had electrodynamics. There still is not such a definitive theory, though candidates have been proposed, including string theory. Nevertheless, every major development that has taken place in the fifty-odd years since Feynman began his work in this area, involving a line of scientists from Feynman to Weinberg to Stephen Hawking and beyond, has built on his approach and on the specific tools he developed along the way.

Here are a few examples:

(1) *Black Holes and Hawking Radiation*: Black holes have remained perhaps the biggest theoretical challenge to

physicists trying to understand the nature of gravity, and they have produced the biggest surprises. While suggestive observational evidence has accumulated in the past forty years of the existence of massive black hole–like objects in the cosmos, from the engines of energetic quasars to million and billion solar mass objects at the centers of galaxies, including our own, the detailed nature of quantum processes that operate in the final stages of black hole collapse has produced surprises and controversy. The biggest surprise came in 1972, when Stephen Hawking explored the detailed quantum mechanical processes that might occur near the event horizon of a black hole, and discovered that these would cause black holes to radiate energy in the form of all types of elementary particles, including gravitons, as if the black hole were hot, at a temperature inversely proportional to its mass. The form of this thermal radiation would be essentially independent of the identity of whatever collapsed to form the black hole, and would cause the black hole to lose mass and perhaps eventually evaporate completely. This result, which is based on the type of approximation Feynman first used to explore the quantum mechanics of gravity—namely, approximating the background space as fixed and approximately flat, and considering quantum fields, including gravitons, propagating in this space—not only flew in the face of commonsense classical thinking but also presented major challenges to our understanding of quantum mechanics in the presence of gravity. What is the source of this finite temperature? What happens to the information that falls down the black hole if the black hole eventually radiates away? What about the singularity at the center of the black hole, where conventional quantum field theory breaks down? These major conceptual and mathematical problems have driven the work of the greatest theoretical minds in physics over the past forty years.

(2) *String Theory and Beyond*: In an effort to tame the infinities of quantum gravity, scientists discovered in the 1960s that if one considers the quantum mechanics of a loop of vibrating string, there is naturally one type of vibration that would be appropriate to describe a massless spin 2 excitation. This led to the recognition, using precisely the results of Feynman described earlier, that Einstein's general relativity might naturally arise in a fundamental quantum theory that incorporated such stringlike excitations. This recognition, in turn, suggested the possibility that such a theory might be a true quantum theory of gravity, in which all of the infinities that Feynman had exposed in his exploration of gravity as a quantum field theory might be tamed. In 1984 several candidate string theories in which all such infinities might disappear were proposed, producing the biggest explosion of theoretical excitement that physics had witnessed since perhaps the development of quantum mechanics itself.

As exciting as this possibility was, however, there was also a minor complication. In order to allow the mathematical possibility of a self-consistent quantum theory of gravity without infinities, the underlying stringlike excitations cannot exist in merely four dimensions. They must "vibrate" in at least ten or eleven dimensions.

How could such a theory be consistent with the four-dimensional world we experience? What would happen to the six or seven extra dimensions? How could one develop mathematical techniques to treat them consistently and still explore phenomena in the world we experience? How could one develop physical mechanisms to hide the extra dimensions? Finally, and perhaps most important, if gravity arose naturally in these theories, in the spirit of Feynman, could the other particles and forces we experience also naturally arise within the same framework?

These became the central theoretical issues that have been explored in the past twenty-five years, and the results have been mixed at best. Fascinating mathematical theorems that have been developed have given exciting new insights into how to understand seemingly different quantum theories as manifestations of the same underlying physics—something that falls precisely within what Feynman described as the central goal of science—and interesting mathematical results that have been obtained may provide insights into how black holes can radiate thermally, appearing to lose information, and still not violate the central tenets of quantum theory. And finally, string theory, which is based on a new type of Feynman diagram to calculate processes involving the behavior of strings, has allowed theorists to discover new ways to classify Feynman diagrams for normal quantum fields, and allowed physicists to derive analytical results in closed form for processes that would have otherwise involved summing an impossibly large number of Feynman diagrams were the calculations performed directly.

But with the good comes the bad. As our understanding of string theory developed, it became clear that it was much more complicated than previously imagined, and that strings themselves are probably not the key objects in the theory, but rather higher-dimensional objects called *branes*, making the possible range of predictions of the theory far more complicated to derive. Moreover, while early hopes had sided with the possibility that a single underlying string theory would make unique and unambiguous predictions yielding all of the fundamental physics measured in laboratories today, precisely the opposite has occurred. Almost any possible four-dimensional universe, with any set of laws of physics, might arise in these theories. If this remains true, then rather than producing "theories of everything," they could produce "theories of anything," which, in the spirit of Feynman, would not be theories at all.

Indeed, Feynman lived long enough to witness the major string revolution of the 1980s and the hype that went along with it. His natural skepticism of grand claims was not swayed. As he put it at the time, "My feeling has been—and I could be wrong—that there's more than one way to skin a cat. I don't think that there's only one way to get rid of the infinities. The fact that a theory gets rid of infinities is to me not a sufficient reason to believe its uniqueness." He also understood, as he had so clearly expressed at the beginning of his 1963 paper on the subject, that any effort to understand quantum gravity suffered from the handicap that any predictions, even in a theory that made clear predictions, might be well beyond the range of experimentation. The lack of predictiveness, combined with the remarkable hubris of string theorists, even with a manifest lack of empirical evidence, motivated him to say, in exasperation, "String theorists don't make predictions, they make excuses!" Or, expressing his frustration in terms of the other key factor that for him defined a successful scientific theory,

> I don't like that they're not calculating anything. I don't like that they don't check their ideas. I don't like that for anything that disagrees with an experiment, they cook up an explanation—a fix-up to say, "Well, it might be true." For example, the theory requires ten dimensions. Well, maybe there's a way of wrapping up six of the dimensions. Yes, that's all possible mathematically, but why not seven? When they write their equation, the equation should decide how many of these things get wrapped up, not the desire to agree with experiment. In other words, there's no reason whatsoever in superstring theory that it isn't eight out of the ten dimensions that get wrapped up and that the result is only two dimensions, which would be completely in disagreement with experience. So the fact that it might disagree with experience is very tenuous, it doesn't produce anything; it has to be excused most of the time. It doesn't look right.

The very issues that aroused Feynman's concerns, expressed more than twenty years ago, have, if anything, been magnified since then. Of course, Feynman was skeptical of all new proposals, including some that turned out to be right. Only time, and a lot more theoretical work, or some new experimental results, will determine whether in this case his intuition was correct.

(3) *Path Integrals in Quantum Gravity and "Quantum Cosmology"*: The conventional picture of quantum mechanics suffers, as I have described, from the problem that it treats space and time differently. It defines the wave function of a system at a specific time and then gives rules for evolving the wave function with time.

However, a basic tenet of general relativity is that such a distinction between space and time is, in some sense, arbitrary. One can choose different coordinate systems, where one person's space is another's time, and the physical results one derives should be independent of this arbitrary separation. This issue becomes particularly important in cases where space is strongly curved—that is, where the gravitational field is strong. As long as gravity is so weak that one can approximate space as being flat, then one can follow the prescription Feynman developed for treating gravity as a small perturbation, and gravitational effects as being primarily due to the exchange of single gravitons moving in a fixed background space. But in the case where gravity is strong, space and time become smeared-out quantum variables, and a rigid separation into a background space and time in which phenomena can evolve becomes problematic, to say the least.

The path-integral formulation of quantum mechanics does not require such a separation. One sums over all of the possibilities for all of the relevant physical quantities, and over all of the paths without requiring a separation of space and time. Moreover, in the case of gravity, where the relevant quantity involves the geometry of space, then one must sum over all of the possible geometries. Feynman's method gives a prescription for doing this, but it is not at all clear that the remaining picture could be handled by the conventional formulation of quantum mechanics.

The path-integral approach has already been applied, most strongly by Stephen Hawking (and later Sidney Coleman and others), to develop a quantum mechanics of the entire universe, where in the path integral one sums over various possible intermediate universes in which strange new topologies are possible, involving baby universes and wormholes. This approach to treating an entire universe quantum mechanically is called quantum cosmology, and involves a host of new and difficult issues, including how to interpret a quantum system with no external observers, and whether the dynamics of the system can determine its own initial conditions, rather than have them imposed by an outside experimenter.

Clearly the field is in its infancy, especially without a well-defined understanding of quantum gravity. But as Murray Gell-Mann lovingly hoped in an essay written after Feynman's death—knowing of Feynman's great desire to discover new laws and not merely reformulate existing ones, as he had feared his approach to QED had done—it could be that Feynman's path-integral formalism is not just a different but equivalent way of formulating quantum mechanics, but rather the only truly fundamental way. As Gell-Mann put it, "Thus, it would have pleased Richard to know that there are now some indications that his PhD dissertation may have involved a really basic advance in physical theory and not just a formal development. The path integral formulation of quantum mechanics may be more fundamental than the conventional one, in that there is a crucial domain where it may apply and the conventional formulation may fail. That domain is quantum cosmology. . . . For Richard's sake (and Dirac's too), I would rather like it to turn out that the path integral method is the real foundation of quantum mechanics, and thus of physical theory."

(4) *Cosmology, Flatness, and Gravitational Waves*: I have saved for last the most concrete, and perhaps least philosophically profound, implication of Feynman's work, because it allows for the possibility of calculations that might be directly compared to experimental data—without which he viewed theoretical efforts as impotent.

Amazingly, Feynman did his work at a time when almost everything scientists now know about the universe on its largest scales was not yet known. Yet his intuition in a number of key areas was right, with one exception, and experiments at the forefront of observational cosmology may soon provide the first direct evidence that his picture of gravitons as the fundamental quanta of the gravitational field is correct.

Feynman realized early on the possibility that the total energy of a system of particles might be precisely zero. As strange as this may sound, it is possible because while it takes positive energy to create particles from nothing, their net gravitational attraction afterward can imply that they have a negative "gravitational potential energy"—namely, that because it takes work to pull them apart to overcome their gravitational attraction, the net energy lost after they are created and are then attracted together might exactly compensate for the positive energy it took to create them. As Feynman put it in his lectures on gravitation, "It is exciting to think that it costs *nothing* to create a new particle."

It is a small step from this, perhaps, to suggest that the total energy of the entire universe might be precisely zero. Such a universe with total energy equal to zero is attractive, because it allows for a universe that began from nothing. All matter and energy we might see could have arisen from a quantum mechanical fluctuation (including a gravitational quantum mechanical fluctuation in space itself). While Feynman speculated on this possibility, the current best model for the evolution of the universe, called inflation, is based on this very idea. The originator of the idea of inflation, Alan Guth, has said that in this case the universe is the ultimate example of a "free lunch."

Interestingly, a universe with zero total gravitational energy is spatially flat—that is, on large scales it behaves like a normal Euclidean space where light travels in straight lines. There is now very good evidence that the universe is flat by direct measurements of its geometry on large scales, one of the most exciting developments in cosmology in recent times. As early as 1963, however, Feynman suggested this was likely to be the case because the fact that gravitationally bound galaxies and clusters of galaxies—the largest bound objects in the universe, tens of millions of light-years across—do exist implied that the positive kinetic energy of the expansion of the universe was roughly balanced by the negative gravitational potential energy in these systems. He was right.

There was one application of his quantum field theory arguments to gravity where he seemed to have departed from his normal sensible physical intuition, however. In his work on QED he, as well as others, had shown that virtual particles not only exist but also are necessary in order to understand the properties of atoms. Thus, empty space is not empty but is a boiling brew of virtual particles. The laws of quantum mechanics tell us that the smaller the scale one wants to consider, the higher the energy the virtual particles that can briefly exist can have. Feynman once referenced this by saying that in the space in the closed palm of a hand, virtual particles existed with enough energy to power our entire civilization. Unfortunately advocates and crackpots have used this statement to express their desire to develop devices that exploit the energy of the vacuum to do precisely this, and solve our energy problems.

What Feynman somehow forgot, and what the Russian physicist Yakov Zel'dovich made clear in 1967, is that all energy gravitates, even the energy of empty space. If empty space had as much energy as Feynman argued, the gravitational forces would be so great as to blow up the earth, because according to general relativity, when energy is put into empty space, the resulting gravitational force is *repulsive*, not attractive. Therefore, the energy of empty space cannot be, on average, orders of magnitude larger than the energy of all matter, or the resulting repulsive force would be so large that galaxies would never have formed.

Nevertheless, Feynman was not completely wrong. The most astounding discovery in the last fifty years, if not longer, has been the discovery that empty space *does* contain energy—far less than Feynman imagined, but enough so that the energy of empty space is currently dominating the expansion of the universe, causing it to accelerate. We currently have no understanding of why this is the case, and why empty space possesses both energy and an amount of energy that is comparable to the total energy contained in all galaxies and matter in the universe. It is probably the biggest mystery in physics, if not all of science.

Feynman's mistake aside, if the idea of inflation—an early period of exponential expansion that would have resulted in a flat universe today and could have generated all of the structures currently observed—is correct, then there is an exciting implication that hearkens back to Feynman's original calculations. If gravitons are elementary particles like photons, then one can calculate that the same quantum mechanical processes which operated during inflation (to eventually produce matter density fluctuations that would have collapsed to form all of the galaxies and clusters we see today) would have also generated a background of gravitons in the early universe, which would today be observable as a background of gravitational waves. This is indeed one of the core predictions of inflation, and an area of physics I have personally been exploring. Most exciting, we may soon be able to detect such a background with satellites that have been sent up to probe the large-scale structure of the universe. If such a background is observed, it will imply that the calculations Feynman performed when he decided to approach gravity like any other field theory allow a prediction that can be compared with observations, meaning at the very least that the apology he offered for thinking about esoteric and undetectable effects in quantum gravity was not necessary.

IT IS APPROPRIATE to conclude this chapter on Feynman's fascination with gravity by once again returning to the apology with which he began his first paper on the subject. Feynman was attracted to quantum gravity because it was off the beaten path. By the same token he realized that that was the case because the only calculations one might perform would result in predictions of effects that were potentially forever unmeasurable, because gravity is so weak. And so as he began his formal discussion of quantum gravitational effects, he stepped back and said, "It is therefore clear that the problem we are working on is not the correct problem; the correct problem is what determines the size of gravitation?"

A more prescient statement could not have been made. The real mystery that has been driving theoretical particle physicists is the question of why gravity is forty orders of magnitude weaker than electromagnetism. Almost all of the current efforts toward unifying forces, including string theory, are directed toward addressing this puzzling and fundamental question about the universe. It is likely that scientists will not have a full and complete understanding of either gravity or the other forces until they are able to answer this question.

This is characteristic of perhaps the single most remarkable feature of Feynman's lasting legacy. Even as he may have failed to resolve the answers to many of nature's fundamental mysteries, he nevertheless unerringly shed light on the very questions that have continued to occupy the forefront of science to this day.

CHAPTER 16

From Top to Bottom

> The game I play is a very interesting one. It's imagination, in a tight strait-jacket.
>
> —RICHARD FEYNMAN

In December of 1959 Feynman gave a lecture at the annual meeting of the American Physical Society, which that year was being held at Caltech. Once again, a desire to strike out in new and uncrowded directions was clearly on his mind, as he began the lecture with a quote I used earlier: "I imagine experimental physicists must often look with envy at men like Kamerlingh Onnes, who discovered a field like low temperature, which seems to be bottomless and in which one can go down and down. Such a man is then a leader and has some temporary monopoly in a scientific adventure." The lecture, published the next year in Caltech's *Engineering and Science* magazine, was titled "There's Plenty of Room at the Bottom." It is a beautiful and fanciful discussion of a whole new world of possibilities that had nothing to do with particle physics or gravity but were firmly grounded in phenomena with direct applications.

In spite of the esoteric field of particle physics that he had chosen to focus upon, Feynman never lost his interest or his fascination with the physics of the world we can see and touch. And so the opportunity to present this lecture represented for Feynman a chance to let his imagination wander over a domain that had always fascinated him, looking for new fodder for his next scientific adventure—somewhere he might have a monopoly. It also represented his own fascination with the remarkable possibilities of physics in two areas that had captured his imagination since he was a child, through to his time at Los Alamos: mechanical devices and computing.

The lecture was a milestone, and has often been reprinted, because it basically outlined a whole new field of technology and science, or rather a set of fields, related to what is now called nanotechnology but not restricted to that realm. Feynman's central point was that in 1959 when most people were thinking of miniaturization, they were being timid—that there was a universe of space between the size of human-scale machines and the size of atoms. By exploiting this space, he imagined we could not only change technology but also open up whole new domains of scientific inquiry that were then beyond the reach of scientists. And these domains were not like quantum gravity but were domains that might be exploited in his own time if people seriously thought about the remarkable universe under their noses. As he put it, "It is a staggeringly small world that is below. In the year 2000, when they look back at this age, they will wonder why it was not until the year 1960 that anybody began seriously to move in this direction."

Feynman began his lecture by saying that some people were impressed by a machine that could write the Lord's prayer on the head of a pin. That was nothing. He envisaged first writing the entire *Encyclopaedia Britannica* on the head of a pin. But, he argued, that was nothing, because one could easily do that with regular printing by simply shrinking the area of each dot used in half-tone printing by a factor of 25,000. As he argued, even then each dot would contain about 1,000 atoms. No problem, he imagined.

But even that was timid, he argued. What about writing all of the information in all of the books in the world? He performed an estimate for doing so that is amusingly similar to one that I did when I tried to consider how much information would be required to store a digital copy of someone for transporting, in *The Physics of Star Trek.* He argued that it would be easy to store one bit of information (that is, a 1 or 0) using, say, a cube of 5 atoms on a side, containing 125 atoms. He also estimated there were about 10^{15} bits of information in all of the books in the world, which at the time he estimated to be about 24 million volumes. In that case, to store all of the information in all of the books in the world would take merely a cube of material less than one-hundredth of an inch on a side—as small as the smallest speck of dust visible to the human eye. Okay, so you get the picture.

Feynman wanted to explore the possibilities of exploiting matter at the atomic level, the range being, as he described, almost unfathomable. Moreover, true to his first love (as he described her in his Nobel lecture), the most exciting thing of all was that once scientists started engineering at this level, they would have to directly confront the realities of quantum mechanics. Instead of building classical machines, they would have to start thinking about quantum machines. Here was a way to merge the quantum universe with the universe of human experience. What could be more exciting?

I was struck, when rereading his lecture, by his remarkable prescience. Many of the possibilities he described have since come to pass, even if not exactly as he imagined, and usually only because he didn't have the necessary data within his straitjacket at the time to imagine properly. Once again, while he might not have directly and personally solved all of the problems, he asked the right questions and isolated the developments that have become the very forefront of technology, a half century later, as well as imagined the principles that might form the basis of technology in the next fifty years. Here are a few examples:

(1) *Writing All of the Books on Earth on a Dust Speck*: How far have we come toward putting all of the books on a speck of dust? When I taught at Yale University in 1988 I bought what was then the largest hard-drive in the physics department. It was 1 gigabyte, and it cost $15,000. Today I own a paperclip-sized memory stick I keep on a keychain. It holds 16 gigabytes and cost me $49. I have a 2-terabyte (or 2,000-gigabyte) portable external hard-drive for my laptop that cost me $150, so I can now buy 2,000 times as much storage for one-one hundredth as much money. Feynman's estimate of 10^{15} bits for all of the books in the world equals about 100 terabytes, or about fifty portable hard-drives. Of course, most of the space in these drives is not for storage, but for the read mechanism, the interfaces with the computer, and power supplies. Moreover, no one has made any effort to miniaturize the storage size beyond that which fits comfort-

ably next to a laptop. We are not yet able to store large amounts of information on atomic-size scales, but we are now only off by a factor of about a thousand.

In 1965, Gordon Moore, the co-founder of Intel, proposed a "law" that the available storage and speed of computers would double about once every twelve months or so. Over the past forty years this goal has been met or exceeded as technology has continued to keep pace with demand. Thus, given the fact that $1,000 = 2^{10}$, we might be a decade away from Feynman's goal, not just of writing but also of reading all of the books in the world on the head of a pin.

(2) *Biology on the Atomic Scale*: As Feynman put it in 1959,

> What are the most central and fundamental problems of biology today? They are questions like: What is the sequence of bases in the DNA? What happens when you have a mutation? How is the base order in the DNA connected to the order of amino acids in the protein? What is the structure of the RNA? Is it single-chain or double-chain, and how is it related in its order of bases to the DNA? What is the organization of the microsomes? How are proteins synthesized? Where does the RNA go? How does it sit? Where do the proteins sit? Where do the amino acids go in? In photosynthesis, where is the chlorophyll? How is it arranged? Where are the carotenoids involved in this thing? What is the system of the conversion of light into chemical energy? It is very

> easy to answer many of these fundamental biological questions; you just look at the thing!

Could he have enumerated more clearly and precisely the frontiers of modern biology? At least three Nobel Prizes have already been awarded for research that allows the sequence of molecular base pairs in DNA to be read at essentially the atomic level. Sequencing of the human genome has been the holy grail of biology, and the ability to determine genetic sequencing has been improving at a rate that far outstrips Moore's law for computers. What cost over a billion dollars to achieve the first time, less than a decade ago, now can be done for several thousand dollars, and it is expected that within the next decade people will be able to sequence their own genome for less than the cost of a good dinner at a restaurant.

Reading out molecules is important, but the real key to advances in biology is determining three-dimensional molecular structures at the atomic scale. Protein structure determines function, and determining how the atomic components of proteins fold up to form a working mechanism is currently one of the hottest topics in molecular biology.

However, as Feynman also anticipated, the ability to probe biological systems at the atomic level is not merely a passive enterprise. At some level, if scientists can read the data, they can write the data—they can build biological molecules from scratch. And if they can build biological molecules, they can ultimately build biological systems—that is, life—from scratch. And if we can build these systems from scratch and understand what makes them function the way they do, then we will be able to design life-forms that don't currently exist on earth, perhaps life-forms that extract carbon dioxide from the atmosphere and make plastic, or algae that produce gasoline. If this sounds far-fetched, it isn't. Biologists like George Church at Harvard, and Craig Venter, whose private company helped first decode the human genome, are working on these challenges right now, and Venter's company recently received $600 million from Exxon for an algae-to-gasoline project.

(3) *Observing and Manipulating Single Atoms*: In 1959 Feynman bemoaned the sorry state of electron microscopy, which itself was a relatively new field. Because electrons are heavy (compared to light, which is massless), the quantum mechanical wavelength of electrons is tiny. This means that while light microscopes are limited by the wavelength of visible light, about 100 to 1,000 times the size of atoms, electrons can be manipulated by magnetic fields to magnify and produce images of far smaller objects from which they scatter. Yet in 1959 the possibility of imaging individual atoms seemed remote, as the energies involved suggested that the systems would have to be disrupted in order to observe them.

How things have changed. Using the very properties of quantum mechanical systems, as Feynman had again anticipated, new microscopes called *scanning-tunnelling* microscopes and *atomic force* microscopes are allowing images of single atoms in molecules to be made. Moreover, Feynman predicted, "The principles of physics, as far as I can see, do not speak against the possibility of maneuvering things atom by atom. It is not an attempt to violate any laws; it is something, in principle, that can be done; but in practice, it has not been done because we are too big." Sure enough "atomic tweezers" have now been developed using techniques similar to those used in the new microscopes, and intense lasers have been created allowing researchers to regularly manipulate and move individual atoms. Once again, three different Nobel Prizes have been awarded for this work.

Scientists can now not only resolve single atoms in space but can also do so in time. Laser technology has allowed the production of laser pulses that last femtoseconds (10^{-15} seconds). This is comparable to the timescale over which chemical reactions between individual molecules occur. By illuminating molecules with such short pulses, researchers hope to observe the sequence of events, at an atomic level, by which these reactions take place.

(4) *Quantum Engineering*: What most excited Feynman about atomic-scale machines and technology was the realization that once one is working at these scales, the strange properties of quantum mechanics become manifest. Understanding this, one might then hope to design materials with specific, and sometimes exotic, quantum mechanical properties. Once again, from his paper:

> What would the properties of materials be if we could really arrange the atoms the way we want them? They would be very interesting to investigate theoretically. I can't see exactly what would happen, but I can

> hardly doubt that when we have some control of the arrangement of things on a small scale we will get an enormously greater range of possible properties that substances can have, and of different things that we can do. When we get to the very, very small world—say circuits of seven atoms—we have a lot of new things that would happen that represent completely new opportunities for design. Atoms on a small scale behave like nothing on a large scale, for they satisfy the laws of quantum mechanics. So, as we go down and fiddle around with the atoms down there, we are working with different laws, and we can expect to do different things. We can manufacture in different ways. We can use, not just circuits, but some system involving the quantized energy levels, or the interactions of quantized spins, etc.

The American Physical Society currently has in excess of forty-five thousand members. Of these professional physicists and students, over half are working in an area called condensed matter physics in which a significant fraction of their effort is devoted not only to understanding the electronic and mechanical properties of materials based on the laws of quantum mechanics, but also to building exotic materials that would do precisely what Feynman had predicted. A host of developments, ranging from high-temperature superconductors, to carbon nanotubes, to more exotic phenomena like quantized resistance and conducting polymers, have been possible as a result, and no less than ten Nobel Prizes in Physics have been awarded for experimental and associated theoretical investigations of the exotic quantum properties of these man-made materials. Just as the transistor—a device that depends on the laws of quantum mechanics for its function and is the basis of essentially every electronic device you use today—completely transformed the world, the technologies of the world of the twenty-first century and beyond will undoubtedly depend on the quantum engineering that is now taking place in research laboratories around the world.

BECAUSE THE INTELLECTUAL excitement of the possibilities he outlined might not be great enough, Feynman decided in 1960 to personally fund two "Feynman" prizes of $1,000. The first would go to "the first guy who can take the information on the page of a book and put it on an area 1/25,000 smaller in linear scale in such manner that it can be read by an electron microscope." The second would go to "the first guy who makes an operating electric motor—a rotating electric motor which can be controlled from the outside and, not counting the lead-in wires, is only 1/64 inch cube." (Alas, Feynman was a product of his time, and in spite of the fact that his sister was a physicist, for him, physicists and engineers were guys.)

In spite of his foresight, Feynman was somewhat behind the time. Much to his surprise (and also disappointment because it really didn't involve any new technology), within a year of the publication of his speech, a gentleman named William McLellan appeared at Feynman's door with a wooden box and a microscope to view his little motor, and claim the second prize. Feynman, who hadn't formally set up the prize structure, nevertheless made good on the

$1,000. However, in a letter to McLellan, he added, "I don't intend to make good on the other one. Since writing the article I've gotten married and bought a house!" He needn't have worried. It took twenty-five years before anyone—in this case, a (male) Stanford University graduate student—successfully followed Feynman's prescription and claimed his prize. By that time $1,000 was not such a significant amount of money.

IN SPITE OF these prizes, and his fascination with practical machines (Feynman continued to consult periodically for companies like Hughes Aircraft throughout his professional life, even at times when he was devoting the major part of his research efforts to things like strange particles and gravity), the idea that most intrigued Feynman in his 1959 lecture, and which he essentially stated as the reason why he was considering these problems in the first place, and the only one he really followed up on later professionally, was the possibility of making a faster, smaller, and totally different kind of computer.

Feynman had long been fascinated by computing machines and computing in general (MIT computer scientist Marvin Minsky has said, incredibly, that Feynman once told him that he was always more interested in computing than physics), an interest reaching an early peak perhaps during his years at Los Alamos, where these activities were vital to the success of the atomic bomb program. He developed totally new algorithms for quickly performing mental estimates of otherwise impenetrably complex quantities and for solving complex differential equations. Recognizing his abilities, even though he was barely out of graduate school, Hans Bethe had made him a group leader on the calculational group, which performed pencil and paper calculations first, then calculations with the hand-operated, clunky machines called Marchant calculators, and finally using new electronic computing machines (which, you may recall, Feynman and his team had removed from boxes and put together before the IBM experts arrived to do so). The group calculated everything from the diffusion of neutrons in a bomb, necessary to determine how much material was needed for a critical mass, to simulating the implosion process vital to the success of a plutonium bomb. He had been nothing short of amazing in every aspect, leading Bethe, with whom he would have mental arithmetic jousting contests, to say he would rather lose any other two physicists than Feynman.

Well before the arrival of the electronic computer, Feynman helped create what might be called the first parallel-processing human computer, presaging the large-scale parallel processors to come. He had already worked his diffusion group into a tightly coordinated team, so that one day when Bethe came in and asked the group to numerically integrate some quantity, Feynman announced, "All right, pencils, calculate!" and everyone flipped their pencils in the air in unison (a trick that he had taught them). This was more than merely play. In the era before the electronic computer, complicated calculations had to be broken up into pieces in order to be performed quickly. Each individual computation was too complicated for any one person or for one Marchant calculator. But he organized a large group, comprising mostly wives of the scientists there, each of whom handled a simple part of a complex calculation and then passed it on to the next person down the line.

Through his experiences, Feynman became intimately familiar with the detailed workings of a computer—how to break problems down so a computer could solve them (turning the computer into an efficient "file clerk," as he called it), and even more interestingly, determining which problems could be solved in a reasonable amount of time and which couldn't. All of this came back to him as he began to think about his process of miniaturization. How small could computers be? What challenges lay ahead, and what gains, in power usage and in the ability to compute, if smaller, more complex computers with more elements could be made? As he put it in 1959, comparing his brain to computers then extant,

> The number of elements in this bone box of mine are enormously greater than the number of elements in our "wonderful" computers. But our mechanical computers are too big; the elements in this box are microscopic. I want to make some that are submicroscopic. . . . If we wanted to make a computer that had all these marvelous extra qualitative abilities, we would have to make it, perhaps, the size of the Pentagon. This has several disadvantages. First, it requires too much material; there may not be enough germanium in the world for all the transistors which would have to be put into this enormous thing. There is also the problem of heat generation and power consumption. . . . But an even more practical difficulty is that the computer would be limited to a certain speed. Because of its large size, there is finite time required to get the information from one place to

another. The information cannot go any faster than the speed of light—so, ultimately, when our computers get faster and faster and more and more elaborate, we will have to make them smaller and smaller.

While Feynman outlined in his 1959 lecture the intellectual challenges and opportunities that led to so many future developments, this last question was the only one he seriously returned to in any detail, and in surprising directions that combined a number of the different possibilities he mentioned in his talk. It took him over twenty years to do so, however. The motivation for his return arose in part from his interest in his son, Carl. By the late 1970s Carl had gone to college, to Feynman's alma mater, MIT, and, happily for Feynman, had switched his area of study from philosophy to computer science. Feynman got interested in thinking more about the field his son was working in. He introduced Carl to MIT professor Marvin Minsky, whom he had met in California, and Minsky introduced Carl to a graduate student living in his basement named Danny Hillis. Hillis had the crazy idea to start a company that would build a giant computer with a million separate processors computing in parallel and communicating with each other through a sophisticated routing system. Carl introduced his father to Hillis—actually he suggested Hillis visit his father when he, Hillis, was out in California. Much to Hillis's surprise, Feynman drove two hours to meet him at the airport to learn more about the project, which he immediately labeled as "kooky," meaning that he would think about its possibilites and practicalities. This machine, after all, would be the modern electronic version of the human parallel computer he had created at Los Alamos. This, combined with the fact that his son was involved, made the opportunity irresistible.

In fact, when Hillis actually started the company Thinking Machines, Feynman volunteered to spend the summer of 1983 working in Boston (along with Carl), but he refused to give vague general "advice" based on his scientific expertise, calling that "a bunch of balony," and demanded something "real to do." He ultimately derived a solution for how many computer chips each router needed to communicate with in order to allow a parallel calculation to be successful. What was striking about his solution was that it was not formulated using any of the traditional techniques of computer science, but rather ideas from physics, including thermodynamics and statistical mechanics. And more important, even though he disagreed with the estimates of the other computer engineers at the company, he turned out to be right. (At the same time he showed how their computer could be put to good use to solve physics problems that numerically challenged other machines, including problems involved in simulating configurations of elementary particle physics systems.)

Around this time, in 1981, he also started to think more deeply about the theoretical foundations of computing itself, and he co-taught a course with Caltech colleagues John Hopfield and Carver Mead that covered issues ranging from pattern recognition to the issue of computability itself. The former was something that had always fascinated him, and about which he had created some outlandish and at the time unworkable proposals for computers. Pattern recognition is still beyond the capabilities of most computers, which is why when you log in to some Web sites, to distinguish human users from automated computer viruses or hackers, they present a picture with letters askew and require you to type what you see before you can proceed.

It was this area, the physics of computation, and the related issue, the computation of physics, that ultimately captured Feynman's attention. He produced a series of scientific papers and a book of lecture notes, published posthumously (after some legal wrangling about his estate), from the course he taught on this subject beginning in 1983.

For a while he was fascinated by the notion of cellular automata, which he discussed at length with a young wunderkind student at Caltech, Stephen Wolfram, who later went on to become famous as the creator of the computer mathematics package Mathematica, which has revolutionized much of the way people do numerical and analytical calculations nowadays. Cellular automata are basically a set of discrete objects on an array that can be programmed to obey simple rules in each timestep of a computer process, depending on the state of their nearest neighbors. Even very simple rules can produce incredibly complicated patterns. Feynman was undoubtedly interested in whether the real world might work this way, with very basic and local rules for each spacetime point at its basis, ultimately producing the complexity seen at larger scales.

But not surprisingly, his primary attention turned to issues in computing and quantum mechanics. He asked himself how one might need to change the algorithms for a computer to simulate a quantum mechanical system rather than a classical one. After all, the fundamental physical rules were different. The system in question would need to be treated probabilistically, and as he had shown in his reformulation of the quantum world, in order to appropriately follow its time evolution one needed to calculate the probability amplitudes (and not the probabilities) of many different alternative paths at the same time. Once again, quantum mechanics as he had formulated it naturally begged for a computer that could perform different calculations in parallel, combining the results at the end of the calculation.

His fascinating ruminations on the subject, contained in a series of papers written between 1981 and 1985, led him in a new direction that hearkened back to his 1959 proposal. Instead of using a classical computer to simulate the quantum mechanical work, could one design a computer with elements so small as to be themselves governed by the rules of quantum mechanics, and if so, how would this change the way a computer could compute?

Feynman's interest in this question apparently came from his continuing interest in understanding quantum mechanics. One might think that he, if anyone, understood how quantum mechanics worked, but in the 1981 lecture and paper where he first discussed this, he made a confession that reveals more about his rationale for choosing problems to think about—in this case quantum computers—than it does about his own lack of comfort with quantum mechanics:

> Might I say immediately, so that you know where I really intend to go, that we always have had (secret, secret, close the doors!)—we always have had a great deal of difficulty in understanding the world view that quantum mechan-

> ics represents. At least I do, because I'm an old enough man that I haven't got to the point that this stuff is obvious to me. Okay, I still get nervous with it. And therefore, some of the younger students . . . you know how it always is, every new idea, it takes a generation or two until it becomes obvious that there's no real problem. It has not yet become obvious to me that there's no real problem, but I'm not sure there's no real problem. So that's why I like to investigate things. Can I learn anything from asking this question about computers—about this may-or-may-not-be mystery as to what the world view of quantum mechanics is?

To investigate this very question, Feynman considered whether it was possible to exactly simulate quantum mechanical behavior with a classical computing system that operates just with classical probabilities. The answer has to be no. If it were yes, that would be tantamount to saying that the real quantum mechanical world was mathematically equivalent to a classical world in which some quantities are not measured. In such a world one would be able to determine only the probabilistic outcomes of the variables one could measure because one wouldn't know the value of these "hidden variables." In this case the probability of any observable event would depend on an unknown, the value of the unobserved quantity. While this imaginary world sounds suspiciously like the world of quantum mechanics (and it was the world Albert Einstein hoped we lived in—namely, a sensible classical world where the weird probabilistic nature of quantum mechanics was due only to our ignorance of the fundamental physical parameters of nature), the quantum world is far more weird than that, as John Bell demonstrated in 1964 in a remarkable paper. Like it or not, a quantum world and a classical world can never be equivalent.

Feynman derived a beautiful physical example of Bell's work by showing that if one tried to mock up a classical computer that could produce the exact same probabilities that a quantum system would produce for some observable quantities as the system evolved, then the probability of some other observable quantity would need to be negative. Such negative probabilities make no physical sense. In a very real way, the world of quantum probabilities is larger than anything that can be embedded in a purely classical world.

While providing a nice physical demonstration of why all hidden-variable theories are doomed to failure, Feynman also asked a more interesting question in his paper. Would it be possible to invent a computer that was quantum mechanical in nature? Namely, if the fundamental computer bits were quantum objects, like, say, the spin of an electron, could one then numerically simulate exactly the behavior of any quantum mechanical system, and thus address quantum mechanical simulations that no classical computer could efficiently handle?

His initial answer in the 1982 paper was a resounding "probably." However, he continued to think about this question, spurred on by work by Charles Bennett, a physicist at IBM laboratories who had demonstrated that much of the conventional wisdom about the physics of computing was incorrect. In particular, the assumption was that every time a computer does a computation, it would have to dissipate energy as heat (after all, anyone who has worked with a laptop knows how hot it can get). However, Bennett showed that it is possible, in principle, to perform a computer computation "reversibly." In other words, it is theoretically possible to perform such a calculation and then perform exactly the reverse operations and end up where one began, without any loss of energy to heat.

The question arose, would the quantum mechanical world, with all of its quantum fluctuations, spoil this result? In a paper in 1985, Feynman demonstrated that the answer was no. But in order to do this, he had to come up with a theoretical model for a universal quantum computer—namely, a purely quantum mechanical system whose evolution could be controlled to produce the necessary logical elements that are part of a universal computing system (that is, "and", "not", "or," etc.). He developed a model for such a computer and described how one might operate it in principle, thus concluding, "At any rate, it seems that the laws of physics present no barrier to reducing the size of computers until bits are the size of atoms, and quantum behavior holds dominant sway."

While the general physical question he was addressing was fairly academic, he realized that the possibility of actually building a computer that was so small that the laws of quantum mechanics would govern the behavior of its individual elements might be of real practical interest. Following his statement that the quantum computer he had theoretically sketched out had been designed to mimic classical computers with each logical operation being done sequentially, he added, almost as a throw away, "What can be done, in these reversible quantum systems, to gain the speed available by concurrent operations has not been studied here." The possibility suggested by this single line could easily change our world. Once again, Feynman had suggested an idea that would dominate research developments in an entire field for a generation, even if he himself did not produce the later, seminal results.

The field of quantum computing has become one of the most exciting areas of theoretical and experimental interest, precisely because of Feynman's argument that classical computers could never exactly mimic quantum systems. Quantum systems are much richer, and therefore it is possible that a "quantum computer" could perform new types of computational algorithms which would allow it to efficiently and realistically complete a calculation that might take the biggest classical computer available today longer than the age of the universe.

The key idea is really the simple feature that Feynman so explicitly exploited in Los Alamos, and that the Feynman path-integral formulation of quantum mechanics so explicitly displays. Quantum systems will, by their very nature, explore an infinite number of different paths at the same time. If each path could be made to represent a specific computation, then a quantum system might be nature's perfect parallel processor.

Consider the system Feynman discussed first, a simple quantum mechanical particle with two spin states, which we might label "up" and "down." If we call the up state "1" and the down state "0," then this spin system describes a typical single computational bit of information. However, the important feature of such a quantum mechanical system is that until we measure it to be in the up or down state, quantum mechanics tells us that it has a finite probability of being in either state, which is tantamount to saying that it is really in both states at the same time. This makes a quantum bit, or *qubit*, as it has now become known, much different from a classical bit. If we can find ways of operating on such a qubit without actually measuring it, and therefore forcing it into a specific state, the possibility exists of having a single quantum processor doing more than one computation at the same time.

In 1994, Peter Shor, an applied mathematician at Bell Laboratories, demonstrated the potential power of such a system, and the world took notice. Shor showed that a quantum mechanical computer could efficiently solve a specific mathematical problem that had proved to be impossible for classical computers to solve in less than what was effectively an infinite time. The problem is simply stated: Every number can be written uniquely as the product of prime numbers. For example, 15 is 3×5, 99 is $11 \times 3 \times 3$, 54 is $2 \times 3 \times 3 \times 3$, and so on. As numbers get larger and larger, it becomes exponentially more difficult to determine this unique decomposition. What Shor proved was that an algorithm could be developed for a quantum computer to explore the space of prime factors of any number and derive the correct decomposition.

Why should we care about this rather obscure result? Well, for those of us who have money in a bank, or use credit cards for transactions, or are concerned about the security of the codes used to keep national secrets safe, then this should matter a lot. All modern banking and national security information is encrypted using a simple code that is impossible for any classical computer to break. The encryption is performed using a "key" that is based on knowing the prime factors of a very large number. Unless we know the factors in advance, we cannot break the code using a normal computer because it would take longer than the age of the universe to do so. However, a sufficiently "large" quantum computer could do the job in a manageable time. What *large* means depends on the complexity of the problem, but systems involving a few hundred or thousand qubits would easily be up to the task.

Should we rush out and take our money out of the bank and hide it under the bed, or rush into our survivalist shelter and await the impending invasion following the breakdown of our national security codes? Obviously not. In the first place, in spite of the huge resources being devoted to ongoing experimental efforts, no one has been able to build a quantum computer out of more than a few qubits. The reason is simple. In order for the computer to behave quantum mechanically, the qubits must be carefully isolated from all outside interactions, which would effectively wash out the quantum mechanical information stored in the system—the same reason we behave classically and not quantum mechanically. In most systems what is normally called *quantum coherence*—the preservation of the quantum mechanical configuration of the separate components of the system—is destroyed in a microscopic fraction of a second. Keeping quantum computers "quantum-like" is a major challenge, and no one knows if this will ultimately be practical in an operational sense.

More important than this practical consideration is the fact that the same quantum mechanical principles that allow a quantum computer to obviate the classical limits in solving such problems as prime factorization also would make possible, in principle, the development of new "quantum transmission" algorithms that allow a completely secure transfer of information from point to point. By this I mean that we would be able to determine with absolute certainty whether a snooping third party has intercepted a message.

The explosion of ideas that has pushed the new field of quantum computing from a twinkle in Feynman's eye in 1960 to the forefront of modern science and technology has been vast, too vast to properly describe here. Ultimately these ideas might lead to changes in the way modern industrialized societies are organized. In a practical sense, these research developments might represent some of Feynman's most important intellectual legacies, even if he didn't live long enough to fully appreciate the significance of his suggestions. It never ceases to amaze me how seemingly esoteric speculations by a bold and creative mind can help change the world.

CHAPTER 17

Truth, Beauty, and Freedom

I don't feel frightened by not knowing things, by being lost in a mysterious universe without any purpose, which is the way it really is, as far as I can tell. Possibly. It doesn't frighten me.

—RICHARD FEYNMAN

On October 8, 1967, the *New York Times Magazine* ran a story titled "Two Men in Search of the Quark." The author, Lee Edson, proclaimed, "The men largely responsible for sending scientists on this wild quark chase are two California Institute of Technology physicists named Murray Gell-Mann and Richard Feynman. . . . One California scientist calls the two men 'the hottest properties in theoretical physics today.' "

At the time, the latter statement was justified. The former was not, however. Over the previous six to seven years, following their joint work on the weak interaction, Feynman had steadily decoupled from the rush to make sense of the emerging confusion in particle physics, as the growing zoo of strongly interacting particles coming out of accelerators seemed designed to taunt all of those who, like the sailors in the *Odyssey*, were attracted to the siren's call, only to crash against the rocks. Gell-Mann, on the other hand, had attacked the situation head-on, with every tool at his and his colleagues' disposal and, after a number of struggles and false starts, had finally brought some hope of clarity to the field.

Feynman summarized the situation in a talk at the sixtieth birthday celebration for Hans Bethe at Cornell, echoing at the beginning statements he made in his first paper on liquid helium in the early 1950s: "One of the reasons why I haven't done anything much with the strongly interacting particles is that I thought there wasn't enough information to get a good idea. My good colleague, Professor Gell-Mann, is perpetually proving me incorrect. . . . We suddenly hear the noises of the crackling of the breaking of the nut."

To get a sense of the confusion that reigned in particle physics during the first half of that decade, we need only reflect on the current best approach to understanding quantum gravity and the associated possible "theory of everything." There are many ideas but little data to guide physicists, and the more we follow up on the theoretical proposals, the more confusing the situation seems to be. In the 1960s, to be sure, machines were producing a lot more data, but no one knew where it was leading. Had anyone suggested in 1965 that within a decade we would develop an almost complete theoretical basis for understanding not only the weak but also the strong force, most physicists would have been incredulous.

Gell-Mann had indeed cracked open the nut with a remarkable insight. At a time when many physicists were considering giving up on even the possibility of developing an understanding of particle physics using the techniques that had worked so well with QED, Gell-Mann, in 1961, discovered the importance of group theory, which gave him a mathematical tool to classify the plethora of new elementary particles according to their symmetry properties. Amazingly, all of the different particles seemed to fall within various *multiplets*, as they were called, in which each particle could be transformed into another particle by the application of a symmetry transformation associated with the group. These symmetries are like the rotations I described earlier, which can leave certain figures, like triangles and circles, looking the same. In Gell-Mann's scheme (and as independently discovered by several others around the world), the different particles fell into sets of representations whose properties (charge and strangeness, for example) could be graphed so that they formed the vertices of a polyhedra, and all of the particles in each polyhedra could then be transformed into each other by symmetries, which could effectively rotate the polyhedra in different directions.

The group that Gell-Mann found could classify strongly interacting particles was labeled *SU(3)*. It basically had eight different internal rotations that could connect particles in different polyhedral-like multiplets of various sizes, although the most obvious representation would have eight members. In this way he was able to classify almost all of the known strongly interacting particles. Exhuberant over the success of his classification scheme (although both he and the rest of the community were far from convinced it was right, based on the evidence then at hand), he called it the "eightfold way," not just because of the numerical property of SU(3), but in a typical Gell-Mann–like fashion, because of a saying of the Buddha about the eight ways to achieve nirvana: "Now this, O monks, is noble truth that leads to the cessation of pain; this is the noble Eightfold Way; namely right views, right intention, right speech, right action, right living, right effort, right mindfulness, right concentration."

When Gell-Mann, as well as the Israeli physicist Yuval Ne'eman, classified the particles this way, one set of nine particles could not be so classified. However, it was known that there was a ten-member representation of the SU(3) symmetry group—a so-called decouplet—that they both independently indicated might be an appropriate choice, suggesting the need for another, as yet undiscovered particle. Gell-Mann quickly announced that such a new particle must exist, which he called the omega-minus, and he sketched out, using symmetry arguments, what its expected properties should be so that experimentalists could look for it.

Needless to say, in a search with all of the drama of a screenplay, just as the experimenters were ready to give up, they found Gell-Mann's particle, with precisely the properties he had predicted, including its strangeness, and a mass within 1 percent of his prediction. The eightfold way had not only survived, it had flourished!

The day after the experimental discovery of omega-minus, at the end of January in 1964, a paper by Gell-Mann appeared in the European physics journal *Physics Letters.* He had decided that his outlandish speculation, and yet another new linguistic gem, would never make it past the strident referees of the U.S. journal *Physical Review.*

It had not been lost on Gell-Mann, and others as well, that the *3* in SU(3) might have some physical significance.

The eight-element multiplets in SU(3) actually could be formed by appropriate combinations of three copies of a smaller representation of the symmetry group, called the fundamental representation, containing three elements. Could it be that these three elements corresponded, in some way, to elementary particles?

The problem was that if strongly interacting particles like protons were made up of three sub-constituents, then these sub-constituents would generally have, by comparison, a fractional electric charge. One of the hallmarks of physics, however, was that all observed particles had electric charges that were integral multiples of the charge on the electron and proton (which had equal and opposite charges). No one knew why this was the case, and to some extent we still don't know. But that is what nature seemed to require.

Nevertheless, after a year or so of discussion and trepidation, spurred on by the discovery of a wonderful line in James Joyce's *Finnegans Wake*—"Three quarks for Muster Mark!"—Gell-Mann wrote a short two-page paper proposing that the eightfold way as a fundamental classification scheme for all strongly interacting particles made mathematical sense if the fundamental constituents of this scheme were three different fractionally charged objects, which he called *quarks.*

Gell-Mann was wary of proposing the existence of a whole new set of exotic, and potentially ridiculous, particles, and by that time the conventional wisdom in the community had swayed toward the idea that fundamental particles themselves might be ill-conceived, and that all elementary particles might be made up of combinations of other elementary particles, in what was called a kind of *nuclear democracy.* Therefore, Gell-Mann was careful to argue that these objects, which he called *up*, *down*, and *strange* quarks, might be just mathematical niceties that allowed the accounting to be done efficiently.

Remarkably, a former Caltech graduate student of Feynman's, George Zweig, who was now a postdoctoral researcher at CERN, the European accelerator laboratory, completed an exactly similar proposal, presented in much more detail, at almost the same time. Moreover, Zweig was much more willing to suggest that these new fractionally charged objects, which he called *aces*, might be real. When he saw Gell-Mann's short paper in print, he quickly tried to get his eighty-page paper published by the *Physical Review*. But Gell-Mann had been wiser, and Zweig was never able to get his work published in that staid journal.

Needless to say, with Gell-Mann's incredible line of greatest hits, from *V-A*, to the omega-minus, it was inevitable that quarks would win out over aces. That is not to say, however, that the physics community reacted with enthusiasm to Gell-Mann's proposal. Instead, it received it with all of the excitement of an unwelcome pregnancy. After all, where were the fractionally charged particles? Searches in everything from accelerator data to the inside of oysters turned up nothing. And thus, even after the *New York Times* had canonized quarks in its 1967 article, Gell-Mann was quoted as saying the quark was likely to turn out to be merely "a useful mathematical figment."

So it was in 1967 that Feynman had finally decided to return to his first love, particle physics, to see what interesting problems he could attack. Despite offering complimentary words about Gell-Mann for the *Times* article, he had not shown much enthusiasm for the work that Gell-Mann had done over the past five years to drive his field forward. He had been highly skeptical of the omega-minus discovery, and quarks had seemed uninteresting—so uninteresting that when his own former student Zweig had proposed aces, Feynman had shown no enthusiasm whatsoever for that idea either. He found the effort of theorists to seek comfort in the language of group theory too much like a crutch that replaced actual understanding. He described how physicists would repeat themselves using the language of mathematics like "simple baby talk, like boo-boo."

While we might suspect that Feynman's reactions were tinged with envy, it is more likely that his natural skepticism was combined with his essential disinterest with what other theorists were thinking. He had thus far held true to the idea that the strong interaction data was too confusing to allow productive theoretical explanation, and he had avoided all of the failed theoretical fads of the 1960s, including the idea of nuclear democracy and its opposition to fundamental particles. His joy was solving problems, and solving them himself. As he said at the time, following a dictum "DISREGARD" that he gave to himself after winning the Nobel Prize, "I have only to explain the regularities of nature—I don't have to explain the methods of my friends." Nevertheless, he had started once again to teach a course on particle physics, and that meant catching up on the field. For Feynman, that meant catching up on the minutiae of the experimental data.

It turned out to be the right time to do so. A new particle accelerator had come online in Northern California, near Stanford, and thus not too far away from Caltech. This new accelerator was based on a different technique for exploring strongly interacting particles. Instead of smashing these particles together and seeing what happened, the SLAC machine, as it became known, accelerated electrons on a two-mile-long track and smashed them into nuclei. Since electrons don't feel the strong force, scientists could interpret their collisions more easily, without the uncertainties of the strong interaction. In this way, they hoped to probe the nucleus just as Ernest Rutherford had done seventy-five years earlier when he discovered the existence of the nucleus by shooting alpha particles at atoms. In the summer of 1968, Feynman decided to visit SLAC during a trip to visit his sister and discover for himself what was happening.

Feynman had already been thinking of how to make sense of the experimental data regarding strongly interacting particles, and I expect that he was influenced by his work on liquid helium. Remember that he had tried to understand how a dense system of atoms and electrons in a liquid could behave at low temperatures as if the atoms were not interacting with each other.

A somewhat similar behavior was suggested by the results of early experiments involving the complex scattering of strongly interacting particles off of each other. In spite of his hesitation to explain the data with some fundamental theory, or maybe because of it, Feynman realized he could explain some general features without recourse to any specific detailed theoretical model. One of the implications of the experimental results was that the collisions mostly took place on the scale of the particles involved, like protons, and not on smaller scales. He reasoned that if the protons had internal constituents, these constituents could not be interacting strongly with each other on smaller scales, or that would have been manifest in the data. Therefore, one could choose to picture strongly interacting particles, or *hadrons*, as they were called, with a simple toy model: a box full of constituents, which he called *partons*, that didn't interact strongly on small scales but were somehow constrained to remain within the hadrons.

The idea was what we call *phenomenological*—namely, it was just a way to make sense of the data, to see if one could probe for regularities in the morass, in order to get some clues of the underlying physics, just as Feynman's picture of liquid helium had done. Of course, Feynman was aware of Gell-Mann's quarks and Zweig's aces, but he was not trying to produce some grand fundamental understanding of hadrons. Rather he wanted to understand how to extract useful information from experiments and so he made no attempt to connect his parton picture to their particles.

Feynman recognized the limitations of his picture, and its distinction from normal model building. As he said in his first paper on the subject, "These suggestions arose in theoretical studies from several directions and do not represent the result of consideration of any one model. They are an extraction of those features which relativity and quantum mechanics and some empirical facts imply almost independently of any model."

In any case, Feynman's picture allowed him to consider a process most physicists, who were trying to explain the data with some fundamental model, had avoided. These others had focused on the simplest of all possibilities, where two particles entered a collision volume and two particles exited the region. Feynman however, realized his simple picture would allow him to explore more complicated processes. In these processes, if experimenters banged hadrons together head-on with enough energy so that a lot of particles were produced, they could hope to measure the detailed energies and momenta of at most a few of the outgoing particles. One might think that in this case they would not get much useful information. But Feynman argued, motivated by his parton picture, that these processes, which he called *inclusive processes*, might actually be worth thinking about.

He realized that at very high energies, the effects of relativity would cause each particle, in the frame of the other particle, to look like a pancake, because lengths along the direction of motion are contracted. Moreover, the effects of time dilation would mean the sideways motions of individual partons around the pancake would appear to be slowed to a standstill. Thus, each hadron would look, to the other hadrons, like a collection of pointlike particles at rest inside a pancake. Then, assuming that the subsequent collision would involve one of the partons from each pancake colliding, with the rest simply passing through one another, physicists could make sense of inclusive processes in which only one outgoing particle in the collision is measured in detail and the rest fly off with only general features of their distribution recorded. Feynman suggested that if this picture of the collision was correct, certain measured quantities, like the momentum of the outgoing particle measured in the direction of the incident beam, should have a simple distribution.

Louis Pasteur is reputed to have said, "Fortune favors the prepared mind." Feynman's mind was well prepared when he visited SLAC in 1968. The experimentalists there had been analyzing their data, the first data on high-energy electrons scattering on proton targets, producing a huge spray of outgoing stuff, according to the suggestion of a young theorist there, James Bjorken, known universally as "BJ." Bjorken is a determined, mild-mannered, brilliant theorist who often speaks in a language that is unfamiliar, but whose conclusions are worth listening to. So it was at SLAC at the time.

Using detailed ideas from field theory, many of which originated with Gell-Mann, Bjorken had shown in 1967 that if experimenters measured merely the properties of the outgoing electrons in these collisions, they would find regularities in their distribution that would be very different if the proton was composed of pointlike constituents than if it wasn't. He called these regularities *scaling properties.*

While the experimentalists involved in the SLAC experiments didn't really understand the detailed theoretical justification for Bjorken's scaling hypothesis, his suggestions did provide one useful way to analyze their data, so they did. And lo and behold, the data agreed with his predictions. Such agreement, however, did not guarantee that Bjorken's somewhat obscure suggestion was correct. Perhaps other mechanisms could produce the same effects.

When Feynman visited SLAC, Bjorken was out of town, and Feynman talked directly to the experimentalists, who, needless to say, gave him a better understanding of the results than why or how Bjorken had derived them. Having already thought about the more complicated hadron-hadron collisions, Feynman realized the electron-proton collisions might be easier to analyze, and the observed scaling might have a simple physical explanation in terms of partons.

That evening, he had an epiphany after going to a topless bar for motivation (there remains some dispute about this), and back in his hotel room he was able to demonstrate that the scaling behavior indeed had a simple explanation: in the reference frame in which the proton looked like a pancake to the incoming electrons, if the electrons bounced off individual partons, each of which was essentially independent, then the scaling function that Bjorken had derived could be understood as simply the probability of finding a parton of a given momentum inside of the proton, weighted by the square of the electric charge on that parton.

This was an explanation the experimentalists could understand, and when Bjorken returned from mountain climbing to SLAC, Feynman was still there and he sought out Bjorken to ask him a host of questions about what he knew and didn't know. Bjorken most vividly remembers the language Feynman used, and how different it was from the way he had thought about things. As he later put it, "It was an easy, seductive language that everyone could understand. It took no time at all for the parton model bandwagon to get rolling."

Needless to say, Feynman was both satisfied and thrilled by the ability of his simple picture to explain the new data. He and Bjorken also realized that other probes of protons could be used to obtain complementary information on the structure of protons by using incident particles that interact not electromagnetically with partons, but via the weak interaction—namely, neutrinos. Feynman was once again at the center of activity in the field, and by the time he published his first paper on the idea, several years after the fact, the analysis of *deep inelastic scattering*, as it had become called, was where all of the action was being focused.

Of course, the central questions then became, Were partons real? and, if so, Were partons quarks? Feynman recognized that the first question was difficult to answer completely, given the utter simplicity of his model and the likelihood that the actual physical phenomena might be more complicated. Years later, in a book on the subject, he stated his concerns explicitly: "It should be noted that even if our house of cards survives and proves to be right we have not thereby proved the existence of partons. . . . Partons would have been a useful psychological guide as to what relations to expect—and if they continue to serve this way to produce other valid expectations they would of course begin to become 'real,' possibly as real as any other theoretical structure invented to describe nature."

As for the second question, that was doubly difficult. First, it took some time, in this climate where the general theoretical bias was against fundamental particles, before people were willing to seriously consider it, and second, even if the partons did represent quarks, why weren't they knocked free, for all to see, emerging from the high-energy collisions?

Over time, however, using the formalism that Feynman had developed, physicists were able to extract the properties of the partons from the data, and lo and behold, the fractional charges on these objects became manifest. By the early 1970s Feynman had become convinced that the partons had all of the properties of Gell-Mann's hypothetical quarks (and Zweig's aces), though he continued to talk in parton language (perhaps to annoy Gell-Mann). Gell-Mann, for his part, deflected criticism that he had not been willing to believe in the reality of quarks by making fun of Feynman's simplified picture. Ultimately, because quarks came from a fundamental model, the physics world moved during the 1970s from the parton picture of protons to the quark picture.

But where were the quarks to be found? Why were they hiding inside of protons, and not found lurking anywhere else? And why were they behaving like free particles inside the proton when the strong interaction that governed the collisions of protons with each other, and hence quarks with each other, was the strongest force known in nature?

Remarkably, within a period of five years, not only were these questions about the strong force essentially answered, but theorists had also developed a fundamental understanding of the nature of the weak force as well. A decade after the mess had begun, three of the four known forces in nature were essentially understood. Perhaps the most significant, and still probably one of the less publicly heralded, theoretical revolutions in the history of our fundamental understanding of nature had been largely completed. The experimentalists at SLAC who had discovered scaling, and hence quarks, won the Nobel Prize in 1990, and the theorists who developed our current "standard model" of the weak and strong forces won Nobel Prizes in 1979, 1997, and 2004.

REMARKABLY, FEYNMAN'S WORK, both during this period and during the previous five years, largely and directly helped make this revolution possible. In the process, without aiming to, and perhaps without his ever fully appreciating the consequences, Feynman's work contributed to a new understanding of the very nature of scientific truth. This in turn implied that his own work on QED was not a kluge but provided a fundamental new physical understanding of why sensible theories of nature on scales we can measure produce finite results.

The story of how all this happened begins, coincidentally, with the work Gell-Mann did with his colleague Francis Low in Illinois in 1953–54. Their paper, which had impressed Feynman when Gell-Mann first visited Caltech, concluded that the effective magnitude of the electric charge on the electron would vary with size, getting larger as one moves closer in and penetrates the cloud of virtual electron-positron pairs that were shielding the charge.

A little bit farther east, in the summer of 1954 Frank Yang and his office mate Robert Mills, at Brookhaven Laboratory on Long Island, motivated by the success of QED as an explanation of nature, published a paper in which they postulated a possible generalization of the theory that they thought might be appropriate for understanding strong nuclear forces.

In QED the electromagnetic force is propagated by the exchange of massless particles, photons. The form of the equations for the electromagnetic interaction is strongly restricted by a symmetry called gauge symmetry, which essentially ensures that the photon is massless, and the interaction is therefore long range, as I have previously described. Note that in electromagnetism, photons couple to electric charges, and photons themselves are electrically neutral.

Yang and Mills suggested a more complicated version of gauge invariance in which many different types of "photons" could be exchanged between many types of "charges," and some of the photons could themselves be charged, which means they would interact with themselves and other photons. The symmetries of these new Yang-Mills equations, as they became known, were both fascinating and suggestive. The strong force didn't seem to distinguish between protons and neutrons, for example, so inventing a symmetry between them, as well as a charged photon-like particle that could somehow couple to and convert one into the other made some physical sense. Moreover, the success in removing infinities in QED depended crucially on the gauge symmetry of that theory, so using it as a basis for the new theory made sense.

The problem was that the gauge symmetry of the new equations would in general require the new photons to be massless, but because the strong interactions are short range, operating only on nuclear scales, in practice they would have to be massive. How exactly this might happen they had no notion, so their paper was not really a model but more an idea.

In spite of these problems, aficionados, like Julian Schwinger and Murray Gell-Mann, continued to return to the idea of Yang-Mills theories during the 1950s and 1960s because they felt their mathematical structure might provide the key to understanding either the weak or the strong force, or both. Interestingly, the group symmetries of Yang-Mills theory could be expressed using the same kind of group theory language that Gell-Mann later used as a classification scheme for the strongly interacting particles.

Schwinger assigned his graduate student Sheldon Glashow the task of thinking about what kind of group structure and what kind of Yang-Mills theory might describe the symmetries associated with the weak interaction. In 1961 Glashow not only found a candidate symmetry, but also showed rather remarkably that it could be combined with the gauge symmetry in QED to produce a model in which both the weak interaction and the electromagnetic interaction arose from the same set of gauge symmetries, and that in this model the photon of QED would be accompanied by three other *gauge bosons*, as the new type of photons became known. The problem was that, once again, the weak interaction was short range while electromagnetism was long range, and Glashow didn't explain how this difference could be accommodated. The moment one gave masses to the new particles, the gauge symmetry, and with it the beauty and potential mathematical consistency, of the model would disappear.

Part of the problem was that no one really knew how to convert Yang-Mills theories into fully consistent quantum field theories like QED. The mathematics was more cumbersome, and the motivation wasn't there to embark on such a task. Enter Richard Feynman. When he first started working on gravity as a quantum theory, the mathematical problems were so difficult that he turned to Gell-Mann for advice. Gell-Mann suggested that he first solve a simpler problem. He told Feynman about Yang-Mills theories and argued that the symmetries inherent in these theories were very similar to, but less intimidating than, those associated with the theory of general relativity.

Feynman took Gell-Mann's advice and analyzed the

quantum properties of Yang-Mills theories and made a number of seminal discoveries, which he wrote up in detail only years later. In particular, he discovered that to get consistent Feynman rules for the quantum theory, one had to add a fictional particle to internal loops to make the probabilities work out correctly. Later two Russian physicists, Ludvig Faddeev and Victor Popov, rediscovered this, and the particles are now called Faddeev-Popov ghost bosons. Moreover, Feynman also discovered a new general theorem about Feynman diagrams in quantum field theories, relating diagrams with internal virtual particle loops to those without such loops.

Feynman's methods for understanding quantized Yang-Mills theories turned out to be of crucial importance for the major developments in physics at the end of the decade. First, Steven Weinberg rediscovered Glashow's model for *electroweak unification*, as it was called, in the context of a specific and more realistic Yang-Mills theory where the weak bosons could, in principle, start out with zero mass—preserving the gauge symmetry—and their mass could arise later, spontaneously, due to the dynamics of the theory.

This was a beautiful potential solution to the problem of finding a theory of the weak interaction. But there remained a problem. Was the theory "renormalizable"? Namely, could one show, as Feynman, Schwinger, and Tomonaga did in QED, that all of the infinities can be efficiently removed in the prediction of physical quantities? In 1972, a young Dutch graduate student, Gerardus 't Hooft and his supervisor, Martinus Veltman, building on Feynman's methods for quantizing these theories, provided the answer: yes. Suddenly the Glashow-Weinberg theory became interesting! Within the next five years experiments began to provide evidence that the theory was correct, including the need for three new heavy-gauge bosons, and in 1984 at CERN the heavy bosons themselves were discovered. All of these developments produced a field day of Nobel Prizes: for Glashow, Weinberg, and Abdus Salam, who had done work similar to theirs, and for 't Hooft and Veltman and the experimentalists who discovered the weak bosons.

Now theorists had wonderful and fundamental theories of the weak and electromagnetic interactions, but the strong interaction remained puzzling. A more complicated Yang-Mills theory, associated with the same symmetry group that Gell-Mann had used to classify quarks, SU(3), seemed promising. In this case the "3" did not correspond to different "flavors" of quarks—that is, up, down, and strange—but rather to some new internal quantum number which became called *color*. This theory seemed to be able to describe the phenomenological features of the way quarks might combine together to form hadrons. In analogy to QED, it was called *quantum chromodynamics*, or *QCD*. However, once again the strong interaction was short range, seeming to require massive bosons.

But more important, how could a strong new force explain the fact that the objects inside protons, whether one called them partons or quarks, act as if they are not interacting? The solution came within a year, and it hearkened back to Gell-Mann and Low's results about the strengthening of the effective value of the electric charge on electrons at small scales.

In 1973, at a time when the stock in quantum field theory seemed to be rising, following the progress in the electroweak theory, a young theorist at Princeton who had been weaned at Berkeley on the nuclear democracy models, which argued that particles and fields were the wrong way to approach the strong interactions, decided to kill the only remaining theory that still had any hope of explaining the strong interaction. David Gross and his brilliant student Frank Wilczek decided to examine the short-distance behavior of Yang-Mills theories, and QCD in particular, with the aim of showing that the effective magnitude of the "color charges" in QCD would, as in QED, appear to increase at short distances due to screening by virtual particles at longer ones. If this were the case there was no hope for such a QCD theory explaining the SLAC scaling results exposed by Feynman and Bjorken. For different reasons, a Harvard graduate student of Sidney Coleman's, David Politzer, was also independently investigating the scaling properties of QCD.

To the surprise of all three scientists, precisely the opposite behavior from what was expected was observed in the resulting equations (once various crucial sign errors were checked and corrected), but only for Yang-Mills theories such as QCD. The effective "color charge" of quarks would not get larger at short distances, but smaller. The theorists dubbed this remarkable and unexpected property, asymptotic freedom. Gross and Wilczek and then Politzer followed up on this discovery with a series of papers in which they adopted precisely the formulation Feynman had developed for making comparisons with the results of the scaling experiments at SLAC. They discovered that not only could QCD explain the scaling, but also, due to the fact that the interactions between quarks were not zero but were nevertheless weaker than they would be without asymptotic freedom, it was possible to calculate corrections to the scaling behavior, which should be observable.

Meanwhile Feynman remained skeptical of all of the excitement about the new results. He had seen theorists get carried away too many times with new grand ideas to jump on any bandwagons. What was particularly interesting was that his skepticism persisted *in spite* of the fact that these new results arose from exploiting the very techniques that he had pioneered, both for understanding scaling experiments and for dealing with Yang-Mills theories.

Eventually—by the mid-1970s—Feynman had become convinced that there was enough merit in these ideas that he began to follow up on them in detail, and with great zest and energy. With a postdoctoral researcher, Rick Field, Feynman calculated a host of potentially physically observable effects in QCD, helping spearhead a new and exciting era of close mutual contact between experiment and theory. It was hard work. The energy scale at which QCD interactions became weak enough that the calculations of the theorists were reliable was somewhat higher than the experimentalists were able to achieve. Therefore, even though tentative confirmation of the predictions of asymptotic freedom were coming in, it took at least another decade—until the mid-1980s, close to the time of Feynman's death—before the theory was fully confirmed. And it took another twenty years before Gross, Wilczek, and Politzer were awarded the Nobel Prize for their work on asymptotic freedom.

During his last years, as much as Feynman remained fascinated with QCD, a part of him continued to resist fully buying into the theory. For while the theory seemed to do a wonderful job explaining the SLAC scaling—and while the subsequent predicted scaling deviations were also observed and indeed all measurements of the strength of the QCD interaction showed it getting weaker at short distances and high energies—on the opposite long-distance scale the theory became unwieldy. This prevented any theoretical test of what would have been the gold standard for Feynman: an explanation of why we don't see any free quarks in nature.

The conventional wisdom is that QCD gets so strong at large distances that the force between quarks remains constant with distance, and therefore it would take, in principle, an infinite amount of energy to pull two quarks fully apart. This expectation has been supported by complex computer calculations, calculations of the type spearheaded by Feynman when he was working on the Connection Machine for Hillis in Boston.

But a computer result was, to Feynman, merely an invitation to understand the physics. As he had learned at the feet of Bethe so many years ago, until he had an analytical understanding of why something happened such that it could produce numbers comparable with experimental data, he didn't trust the equations. And he didn't have that. Until he did, he wasn't willing to lay down his sword.

This was when I first met Richard Feynman, as I described at the beginning of this book. He came to Vancouver and lectured with great excitement on an idea he thought could prove that QCD would be *confining*, as the problem of the "non-observation" of isolated free quarks was called. The problem was too difficult to treat in three dimensions, but he was pretty sure that in two dimensions he could develop an analytical approach that would finally settle the matter in a way that would satisfy him.

Feynman continued to press on hard, through his battle with cancer, first treated in 1979 and then reappearing in 1987, and through the increasing distractions associated with his growing fame, from activities surrounding his best-selling autobiographical books to his stint on the *Challenger* commission (where he personally helped uncover the reason for the tragic space shuttle explosion). But he never lived to see his goal realized. To this very day, while computer calculations have improved tremendously, giving more and more support to the notion of confinement, and while a host of new theoretical techniques have allowed sophisticated new ways of dealing with Yang-Mills theories, no one has come up with a simple and elegant proof that the theory must confine quarks. No one doubts the theory, but the "Feynman test," if one might call it that, has not yet been met.

Feynman's legacy lives on, however, every single day. The only truly efficient and productive techniques for dealing with both Yang-Mills gauge theories and gravity involve Feynman's path-integral formalism. Essentially no other formulation of quantum field theory is used by modern physicists. But more important, the results of path integrals, asymptotic freedom, and the renormalizability of the strong and weak interactions have pointed physicists in a new direction, giving a new understanding of scientific truth in a way that should have made Feynman finally feel proud of the work he did on QED, instead of feeling that he had merely found an elegant way to sweep problems under the rug.

Feynman's path-integral methodology allowed physicists to systematically examine how the predictions of the theory change as one changes the distance scale at which one chooses to alter the theory to remove the effects of higher- and higher-energy virtual particles in order to renormalize the theory. Because in his language quantum theories are formulated by explicitly examining space-time paths, one can "integrate out" (that is, average over) the very small wiggles in paths appropriate to these scales, and thereby consider only paths that no longer have such wiggles.

The physicist Kenneth Wilson, who later won a Nobel Prize, demonstrated that this integrating out means that the resulting theory, the finite theory, is really only an "effective theory," one that is appropriate to describe nature on scales larger than the cutoff scale where small wiggles in paths are integrated out.

Feynman's technique of getting rid of infinities then was not an artificial kluge, but rather physically essential. This is because we now realize we should no longer expect a theory to hold, unaltered, at all energy and distance scales. No one expects QED, the best-tested and most-beloved theory in physics, to remain the appropriate description of nature as the scales get smaller and smaller. Indeed, as Glashow, Weinberg, and Salam demonstrated, at a sufficiently high-energy scale QED merges with the weak interaction to form a new unified theory.

We now understand that *all* physical theories are merely effective theories that describe nature on a certain range of scales. There is no such thing yet as absolute scientific truth, if by that we mean a theory that is valid at all scales at all times. The physical need for renormalization is then simple: the infinite theory—namely, the one where we extrapolate our theory down to arbitrarily small distance scales—is *not* the right theory and the infinities are the sign of this. If we choose to so extrapolate the theory, we are doing so beyond its domain of validity. By cutting off the theory at some small scale, we are simply ignoring the unknown new physics which would inevitably change the theory at these scales. The finite answers we get are meaningful precisely because if we wish to probe phenomena at large distance scales, we *can* ignore this unknown new physics at tiny scales. Sensible, renormalizable theories like QED are insensitive to new physics at distance scales well below those scales where we perform experiments to test the theories.

Feynman's hope that somehow we would be able to solve the infinity problem in QED without renormalization was therefore a misplaced hope. We now know that his picture, which allows us to systematically see how to ignore the things we do not understand, is as good a one as we are likely to get. In short, Feynman did as much as was possible, and far from hiding the problems of field theory, his mathematical fix was much more than that. It truly demonstrated new physical principles that he had always hoped he would one day be responsible for discovering.

This new understanding would have pleased Feynman, not just because it gives new significance to his own early work but because it keeps the mysteries fresh. No currently known theory is the final answer. He would have liked that. As he once said, "People say to me, 'Are you looking for the ultimate laws of physics?' No, I'm not. I'm just looking to find out more about the world. If it turns out there is a simple, ultimate law which explains everything, so be it; that would be very nice to discover. If it turns out it's like an onion, with millions of layers, and we're sick and tired of looking at the layers, then that's the way it is. But whatever way it comes out, it's nature, and she's going to come out the way she is."

At the same time, the remarkable developments of the 1970s made possible by building on Feynman's work led many physicists to strike out in another direction. After the success of electroweak unification and asymptotic freedom, a new possibility arose. After all, as Gell-Mann and Low showed, QED gets stronger at small scales. And as Gross, Wilczek, and Politzer demonstrated, QCD gets weaker at small scales. Maybe if we went to a very small scale, which we estimate might be sixteen orders of magnitude smaller than the size of a proton, and some twelve orders of magnitude smaller than the best current accelerators can probe, all the known forces might become unified in a single theory, with a single strength. This possibility, which Glashow dubbed *grand unification*, became the driving force for particle physics in much of the 1980s, subsumed by an even grander goal when string theory was discovered to allow a possible unification of the three nongravitational forces with gravity.

Feynman, however, remained suspicious. All his life he had fought against reading too much into data, and he had witnessed a host of brilliant, elegant theories fall by the wayside. Moreover, he knew that unless theorists are willing and able to continue to test their ideas against the cold light of experimentation, the possibility for self-delusion remains great. He knew, as he often said, that the easiest person to fool is yourself.

When he railed against pseudo-scientists, alien-abduction "experts," astrologers, and quacks, he tried to remind us that we seem to be hard wired to find that what happens to each of us naturally appears to take on a special significance and meaning, even if it is an accident. We have to guard against this, and the only way to do so is by adhering to the strait-jacket of empirical reality. So, when faced with claims that the end of physics was near and the ultimate laws of physics were right around the corner, Feynman simply uttered, with the wisdom of age, "I've had a lifetime of that . . . a lifetime of people who believe that the answer is just around the corner."

If the remarkable professional life of one of the most remarkable scientists of the twentieth century is to teach us anything, it is that the excitement and hubris that naturally follow from the rare privilege of uncovering even a small slice of nature's hidden mysteries need to be tempered by the realization that however much we have learned, more surprises are in store for us, if we are willing to carry on searching. For a fearless and brilliant adventurer like Richard Feynman, this was the reason for living.

EPILOGUE

Character Is Destiny

> The way I think of what we're doing is, we're exploring—we're trying to find out as much as we can about the world . . . my interest in science is to simply find out more about the world, and the more I find out, the better it is.
>
> —RICHARD FEYNMAN

Richard Feynman died shortly before midnight on Februrary 15, 1988, at the age of sixty-nine. In those few years he had managed to change not only the world, or at least our understanding of it, but also the lives of everyone he met. No one who had the privilege of knowing him was untouched. There was something so unique about him that it was impossible to view him as one viewed others. If it is true that character is destiny, he then seemed born to discover great things, even as his discoveries were the product of unbelievably hard work, boundless energy, and a rigid integrity aligned with a brilliant mind.

It may also be true that as much as he achieved, he could have accomplished much more had he been more willing to listen and learn from those around him, and insist less on discovering absolutely everything for himself. But accomplishment was not his purpose. It was learning about the world. He felt the fun lay in discovering something, for himself, even if everyone else in the world already knew it. Time after time, when he found out that someone else had scooped him in a discovery, his reaction was not one of despair, but rather, "Hey, isn't that great that we got it right?"

Perhaps we can learn the most about a person by the collective reactions of those around him, and so to complete the picture of Richard Feynman, I decided to include some of these reactions that did not make it into the preceding pages, but that might illuminate more fully the remarkable experience of knowing the man, and one or two stories that, for me, capture his essence.

First, consider the experience of a young student, Richard Sherman, who was fortunate enough to spend an afternoon in Feynman's office:

> I can recall one episode that I found particularly awesome. Midway through my first year I was doing research on superconductivity, and one afternoon I went into his office to discuss the results. . . . I started to write equations on the blackboard, and he began to analyze them very rapidly. We were interrupted by a phone call. . . . Feynman immediately switched from superconductivity to some problem in high-energy-particle physics, into the middle of an incredibly complicated calculation that was being performed by somebody else. . . . He talked with that person for maybe five or ten minutes. When he was through, he hung up and continued the discussion on my particular calculations, at exactly the

> point he had left off. . . . The phone range again. This time it was somebody in theoretical solid-state physics, completely unrelated to anything we had been speaking about. But there he was, telling them, "No. No, that's not the way to do it. . . . You need to do it this way. . . ." . . . This sort of thing went on over about three hours—different sorts of technical telephone calls, each time in a completely different field, and involving different types of calculations. . . . It was staggering. I have never seen this kind of thing again.

Or a not-too-different experience related by Danny Hillis, after Feynman started his summer job at Thinking Machines:

> Often, when one of us asked him for advice, he would gruffly refuse with, "That's not my department." I could never figure out just what his department was, but it didn't matter anyway, because he spent most of his time working on these "not my department" problems. . . . More often than not he would come back a few days after his refusal and remark, "I've been thinking about what you asked the other day and it seems to me . . ." . . . But what Richard hated, or at least pretended to hate, was being asked to give advice. So why were people always asking him for it? Because even when Richard didn't understand, he always seemed to understand better than the rest of us. And whatever he understood, he could make others understand as well. Richard made people feel like children do when a grown-up first treats them as adults. He was never afraid to tell the truth.

Solving problems was not a choice for Feynman, it was a necessity, as his college chum Ted Welton realized early on. Feynman couldn't have stopped if he tried, and he didn't try because he was so good at it. Not even a fatal illness could stop him. Consider a story his Caltech colleague David Goodstein told to the filmmaker Christopher Sykes:

> One day Feynman's secretary Helen Tuck called me up to tell me quietly that Dick had cancer and that he would be going into the hospital for an operation the following Friday. . . . This particular Friday, a week before the operation . . . I told him that somebody had found an apparent error in a calculation that we had done . . . and I didn't know what the error was. Would he be willing to spend some time with me to look for it? And he said, "Sure." . . . On Monday morning we met in my office and he sat down and started working. . . . Most of the time I just sat there looking at him, and thinking to myself, "Look at this man. He faces the abyss. He doesn't know whether he is going to live through this week, and here is this really unimportant problem in two-dimensional elastic theory." But he was consumed by it, and he worked on it all day long. . . . Finally, at six o'clock in the evening, we decided that the problem was intractable . . . so we gave up and went home. . . . Two hours later, he called me at home to say that he had solved the problem. He hadn't been able to stop working on it, and finally he had found the solution to this utterly obscure problem . . . he was exhilarated, absolutely walking on air. . . . This was four days before the operation. I think that tells you a little bit about what drove the man to do what he did.

For Feynman, the process was what he loved. It was a release from the tedium of existence. Stephen Wolfram, who created Mathematica, was a young protégé of Feynman's for several years while he was a student at Caltech, and he described something similar:

> It was probably 1982. I'd been at Feynman's house, and our conversation had turned to some kind of unpleasant situation that was going on. I was about to leave. And Feynman stops me and says: "You know, you and I are very lucky. Because whatever else is going on, we've always got our physics." . . . Feynman loved doing physics. I think what he loved most was the process of it. Of calculating. Of figuring things out. . . . It didn't seem to matter to him so much if what came out was big and important. Or esoteric and weird. What mattered to him was the process of finding it. . . . Some scientists (myself probably included) are driven by the ambition to build grand intellectual edifices. I think Feynman—at least in the years I knew him—was much more driven by the pure pleasure of actually doing the science. He seemed to like best to spend his time figuring things out, and calculating. And he was a great calculator. All around perhaps the best human calculator there's ever been. I always found it incredible. He would start with some problem, and fill up pages with calculations. And at the end of it, he would actually get the right answer! But he usually wasn't satisfied with that. Once he'd got the answer, he'd go back and try to figure out why it was obvious.

When Feynman took an interest in something, or someone, that was it. The effect was magnetic. He focused all of his energy, his concentration, and, it seemed, his brilliance on that one thing or person. That is why so many people were so affected when Feynman came to listen to their seminars and remained to ask questions.

Because the reactions of colleagues to Feynman were generally so intense, they tended to reflect not only Feynman's character but also that of the colleagues. For example, I asked David Gross and Frank Wilczek, two very different individuals who discovered asymptotic freedom in QCD, how Feynman had reacted to QCD and their 1973 results. David told me he was irritated that Feynman had not shown enough interest, largely, David felt, because Feynman hadn't derived the result. Later, when I spoke to Frank about the same subject, he told me how honored and surprised he was by the interest Feynman had displayed. He said Feynman was skeptical, but in those early years Frank thought that that was the appropriate response. I suspect they were both right.

The most telling story that captures the Richard Feynman that I have come to know in writing this book, and the principles that guided his life and directed the nature of his physics, was told to me by a friend, Barry Barish, who was Richard's colleague at Caltech for the last twenty years of his life. Barry and Richard lived relatively close by, so they would often see each other. And since they both lived about three miles from campus, they would sometimes walk, rather than drive, to work—sometimes together, sometimes not. One time Richard asked Barry if he had seen a certain house on a certain street and what he thought of it. Barry didn't know the house because, like most of us, he had found a route he favored and took that route to work and back every journey. Richard, he learned, made a point of doing precisely the opposite. He tried never to take the same path twice.

Acknowledgments and Sources

As I indicated in the introduction, one of the reasons why I agreed to write this volume, after the idea was proposed to me by James Atlas, was that it provided me with the opportunity, and motivation, to go back and read, with varying levels of detail, all of Feynman's scientific papers. I knew the experience, as a physicist, would be enlightening and would allow me to better understand the actual course of physics history, instead of the revisionist version that inevitably develops as physicists refine and simplify techniques that were once obscure.

Nevertheless, I make no pretense to have performed any sort of fundamental historical scholarship. While I have pursued some historical investigations in the past, which required me to go to archives and search out letters and other primary source documents, in the case of Richard Feynman almost all of the primary material I have needed has been nicely compiled and is available in published form.

When this is supplemented by two extraordinary books, one focusing primarily on Feynman's life and the other on the detailed physics history of his work on quantum electrodynamics, an interested and technically trained reader can have direct access to almost all of the material I used as a basis for this book.

Outside of these sources, I am grateful to many of my physics colleagues for discussions about their impressions and personal experiences with Feynman. These include, but are not limited to, Sheldon Glashow, Steven Weinberg, Murray Gell-Mann, David Gross, Frank Wilczek, Barry Barish, Marty Block, Danny Hillis, and James Bjorken. In addition, I thank Harsh Mathur for helping, as he often has for me, to act as a preliminary guide to the condensed matter literature, in this case to the work of Feynman in this area.

The major sources of information that interested readers can turn to, and which incidentally provide every Feynman quote one can find in this book, include published primary source material by Feynman and about Feynman. These include, as I have described, a comprehensive technical presentation of not only Feynman's work on QED but also reproductions of all of his major papers, and a wonderful and definitive personal biography of his life. In addition, there are several excellent references including a recent illuminating compilation of Feynman's letters and various compendia of reflections on Feynman by those who knew him, scientists and otherwise:

QED and the Men Who Made It, Sylvan S. Schweber, Princeton University Press, 1994.

Selected Papers of Richard Feynman, Laurie Brown (ed.), World Scientific, 2000.

Genius: The Life and Science of Richard Feynman, James Gleick, Pantheon, 1992.

Perfectly Reasonable Deviations: The Letters of Richard Feynman, M. Feynman (ed.), Basic Books, 2005.

Most of the Good Stuff: Memories of Richard Feynman, Laurie Brown and John Rigden (eds.), Springer Press, 1993 (proceedings of an all-day workshop in 1988 in which key scientists wrote their reflections of Feynman).

No Ordinary Genius: The Illustrated Richard Feynman, Christopher Sykes (ed.), W. W. Norton, 1994.

The Beat of a Different Drum: The Life and Science of Richard Feynman, Jagdish Mehra, Oxford University Press, 1994.

Three useful additional sources include historical studies of physics and other physicists:

Pions to Quarks: Particle Physics in the 1950s, Laurie M. Brown, Max Dresden, Lillian Hoddeson (eds.), Cambridge University Press, 1989.

Strange Beauty: Murray Gell-Mann and the Revolution in the Twentieth Century Physics, G. Johnson, Vintage, 1999.

Drawing Theories Apart: The Dispersion of Feynman Diagrams in Postwar Physics, David Kaiser, University of Chicago Press, 2005.

Finally, useful scientific books by Feynman include:

QED: The Strange Theory of Light and Matter, Princeton University Press, 1985.

The Character of Physical Law, MIT Press, 1965.

The Feynman Lectures on Computation, A. J. G. Hey and R. W. Allen (eds.), Perseus, 2000.

The Feynman Lectures on Gravitation, with F. B. Moringo, and W. G. Wagner; B. Hatfield (ed.), Addison-Wesley, 1995.

Statistical Mechanics: A Set of Lectures, Addison-Wesley, 1981.

Theory of Fundamental Processes, Addison-Wesley, 1961.

Quantum Electrodynamics, Addison-Wesley, 1962.

Quantum Mechanics and Path Integrals, with A. Hibbs, McGraw-Hill, 1965.

The Feynman Lectures on Physics, with R. B. Leighton and M. Sands, Addison-Wesley, 2005.

Nobel Lectures in Physics, 1963–72, Elsevier, 1973.

Elementary Particles and the Laws of Physics: The 1986 Dirac Memorial Lectures, with S. Weinberg, Cambridge University Press, 1987.

The Meaning of It All: Thoughts of a Citizen Scientist, Helix Books, 1998.

Feynman's Thesis: A New Approach to Quantum Theory, Laurie Brown (ed.), World Scientific, 2005.

Index

Page numbers in *italics* refer to illustrations.